Orthopaedic Biomechanics

The Application of Engineering to the Musculoskeletal System

Orthopaedic Biomechanics

The Application of Engineering to the Musculoskeletal System

Victor H. Frankel, M.D., Med. Dr., Docent

Associate Professor of Orthopaedic Surgery and
Biomedical Engineering
Director of the Biomechanics Laboratory
Division of Orthopaedic Surgery
Case Western Reserve University
School of Medicine
Associate Orthopaedist, University Hospitals and Rainbow Hospital
Chief of Orthopaedic Surgery
Veterans Administration Hospital
Cleveland, Ohio

Albert H. Burstein, Ph.D.

Assistant Professor of Orthopaedic Surgery and
Biomedical Engineering
Co-Director of the Biomechanics Laboratory
Division of Orthopaedic Surgery
Case Western Reserve University
School of Medicine
Consultant, Prosthetics and Orthotics
Veterans Administration Hospital
Cleveland, Ohio

220 Illustrations

Lea & Febiger

Philadelphia, 1970

SBN 8121–0090–5

Published in Great Britain by Henry Kimpton, London

Library of Congress Catalog Card Number 77–78537

Printed in the United States of America

Preface

This is a book of engineering principles that apply to the locomotor system. While containing neither a critical review of the literature nor an in-depth treatment of engineering science, it will serve as an introduction for the orthopaedic surgeon, physical therapist, and student of the locomotor apparatus to those areas of engineering science which are useful both in clinical and research activities. The reader who desires to progress further into the area of biomechanics will benefit by first mastering the calculus and then proceeding to study those areas of engineering introduced in this text.

We are grateful to the Hartford Foundation, the Social and Rehabilitation Service of the Department of Health, Education, and Welfare, the Orthopaedic Research and Education Foundation, and the Rainbow Research Fund, all of which supported the research used to illustrate the physical principles and engineering science expounded in this volume. The authors extend their appreciation to Stanley Waine, Director of the Audio-visual Department, Albert Einstein College of Medicine, Yeshiva University, New York, who capably illustrated this text. Some of the work was performed at the Biomechanical Laboratory, Department of Orthopaedic Surgery, University of Uppsala, Sweden, under the stimulation and guidance of Professor Carl Hirsch. We are indebted to Dr. J. E. Milgram for encouragement and helpful criticism. We would also like to thank the orthopaedic residents and surgeons who attended our courses and through their interest demonstrated that it is possible to acquire a knowledge of those aspects of engineering which apply to orthopaedic surgery in the manner presented in this book.

Cleveland, Ohio — VICTOR H. FRANKEL
Gay Head, Mass. — ALBERT H. BURSTEIN

Contents

1. Statics and Dynamics

IN HIS daily practice the orthopaedic surgeon deals with the effects of forces. He may alter force systems by transferring a tendon, by an arthrodesis of a spine, by designing an arch support, or by performing an osteotomy. He deals with the consequences of internal effects of externally applied forces and moments when a fracture is plated or a dislocation is reduced.

The analysis and correction of disorders of the musculoskeletal system are based on an understanding of the workings and properties of the member involved. This includes a knowledge of statics, dynamics, and the strength and behavior of materials.

Statics is the study of the external effects of forces on a body in equilibrium. Dynamics is the study of the action of forces on bodies not in equilibrium. In human locomotor activity, both equilibrium and nonequilibrium phases must be studied. It is therefore necessary to have a precise and thorough knowledge of the nature of forces.

A force may be thought of as either a push or a pull. It may be defined, however, according to its effects upon the body on which it is acting. These effects as stated above are both internal and external. The external effect of a force tends to change the velocity of the body. The internal effect of the force on the body is to cause a state of strain. Chapter 2 will deal with this latter aspect.

A body moving at a constant velocity requires a force to change this velocity. A special case of this situation is the velocity of a stationary body. It has zero velocity, and in order to change its velocity a force must be applied. A body moving at a constant velocity is said to be in equilibrium.

Forces possess four definite characteristics (Fig. 1–1):

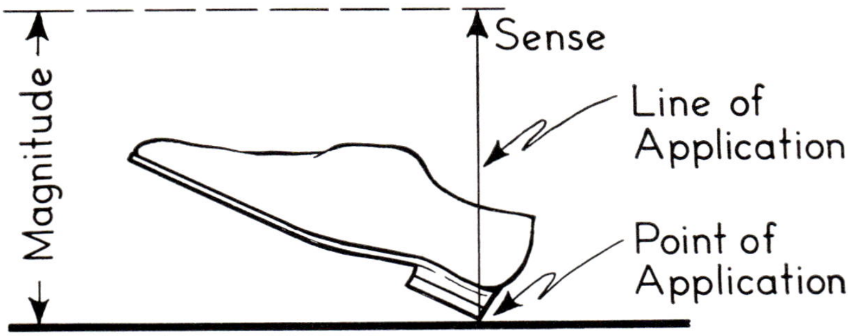

Fig. 1–1. The four vector characteristics of the force of the ground on the heel during heel strike.

1. Magnitude, *i.e.,* kilogram force (kg_f) or pound force (lb_f)
2. Line of application, *i.e.,* vertical, horizontal, or north by west
3. Sense, *i.e.,* upward along a vertical, or downward along a vertical
4. Point of application, *i.e.,* heel of foot, tibial tubercle, or tip of arrow.

The term direction is usually used to include line of application and sense.

Since quantities which possess the above four characteristics are vectors, forces are vectors. They may be studied by vector analysis. It is necessary to understand these four vector characteristics if one wishes to understand the effect of forces on bodies.

The first vector characteristic is magnitude. It is descriptive of the amount of force being applied. By convention this is expressed as kilogram force, abbreviated as kg_f, or pound force, abbreviated lb_f. The magnitude of a force may be measured on a spring scale or on a balance by comparing it to the force produced by a known weight.

Line of application, the second vector property, is a description of the line along which the force is acting. It may be described as being vertical, horizontal, north-south, or by some other geometric description of the orientation of the line. If the force is exerted on a north-south line, it is evident that there are two different senses the force may possess. The force may be pointing north or pointing south.

Sense, the third vector characteristic, chooses between the two possible headings of a vector along a line of application.

The point of application is simply the point of contact between the applied force and the body to which the force is applied. The point of application of a force must, therefore, lie along its line of application. In Figure 1–1, the point of application of the force that the ground exerts against the foot is the heel. This is the point of contact with the ground at the instant the force is being considered. The foot also applies a force to the ground. These forces have equal magnitude, the same line of application, but opposite senses, that is, the ground applies a force upward on the foot and the foot applies a force downward on the ground. Both forces have the same magnitude at the instant of consideration.

Vectors may be manipulated by several established methods. The graphical method is useful in the study of biomechanics. In this method, vectors are drawn to scale and the calculations are performed graphically. A typical traction setup (Fig. 1–2) illustrates this prin-

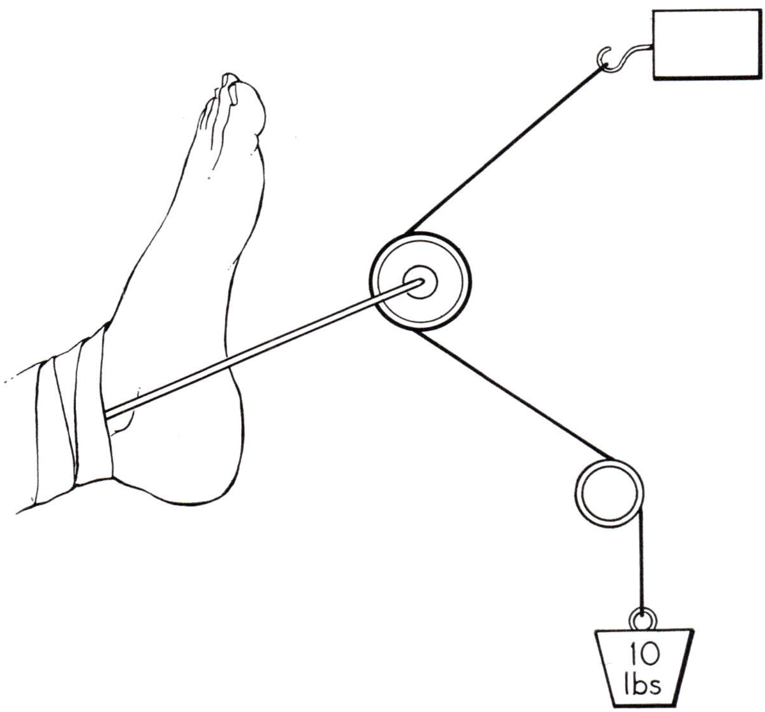

Fig. 1–2. Traction for a fractured femur can be used to illustrate vector properties.

ciple. The rope exerts a 10-pound pull on each side of the pulley. The vectorial sum of the two forces is 18 pounds. This is determined by first connecting the vectors by extending their lines of application. It is proper to extend a vector past its point of application since, in considering only the external effect of a force, the point of application may be varied anywhere along the line of application without changing the external effect. The weight pan in the traction device can be hung higher or lower, and unless it strikes the floor, it will have the same force effect for any of its points of application, as shown in Figure 1–3. The 10-pound lengths are then scaled off, and

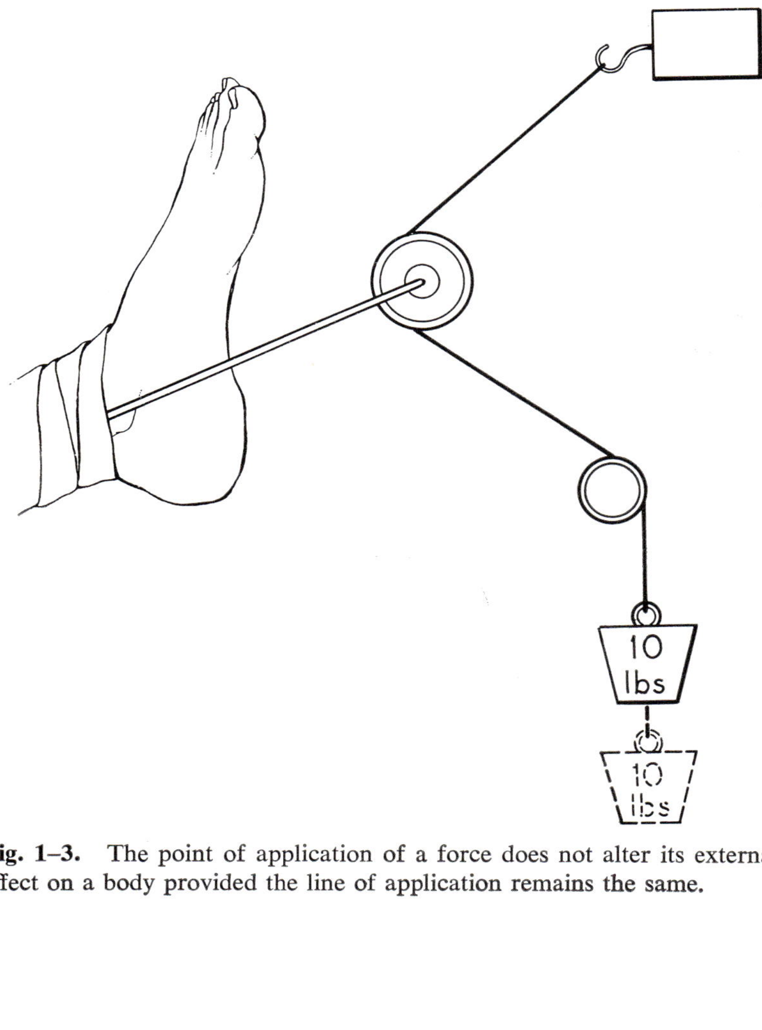

Fig. 1–3. The point of application of a force does not alter its external effect on a body provided the line of application remains the same.

a parallelogram is constructed (Fig. 1–4). The diagonal of the parallelogram is drawn, and this represents the vector sum of the two forces. The diagonal can be measured directly and is found to be equivalent to 18 pounds.

The graphical method presented is called the parallelogram method of addition for vectors. If there are more than two vectors to be added, successive parallelograms must be drawn (Fig. 1–5). This is accomplished by adding any of the two vectors together in the first step and then adding the third vector to the sum of the first two vec-

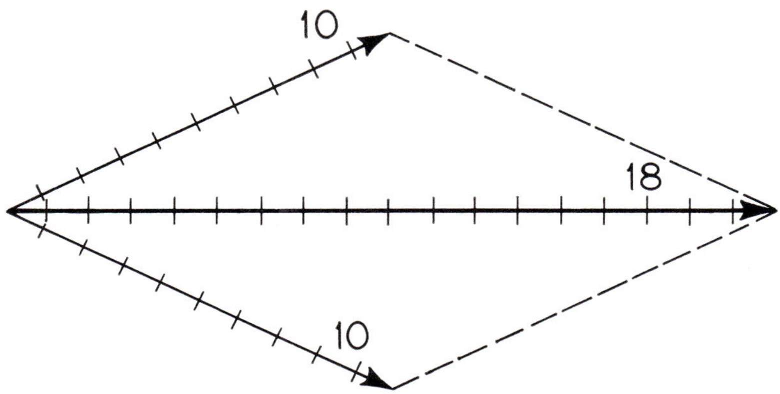

Fig. 1–4. Vector addition of two 10-pound forces.

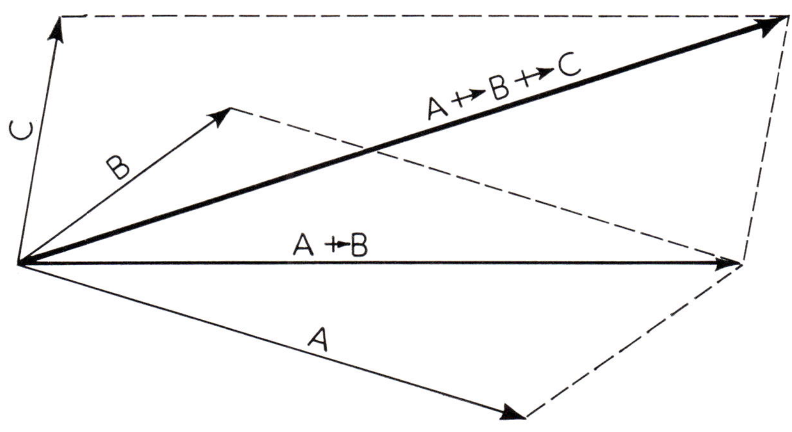

Fig. 1–5. Addition of three vectors.

tors. This sum of the first two vectors is, of course, the diagonal of their parallelogram. Thus, in adding three vectors, two parallelograms must be drawn.

It is important to note that a force has an external effect on the body upon which it acts, not only along the line of application of the force, but also in other directions. When a crutch tip contacts the ground, the direction of the contact force may not be colinear with the axis of the crutch (Fig. 1–6). That is to say, the line of application of the force may not coincide with the shaft of the crutch. The crutch contacts the ground at an angle of 75 degrees while the measured force acts at an angle of 50 degrees to the ground. The force, although acting at an angle of 50 degrees to the ground, still helps to support the weight of the patient. How much of this force is effectively supporting the weight of the patient? The portion of the force that is useful in supporting the weight of the patient is the component of the force acting in the vertical direction This portion of the force, or component of the force, may be determined by a simple graphical

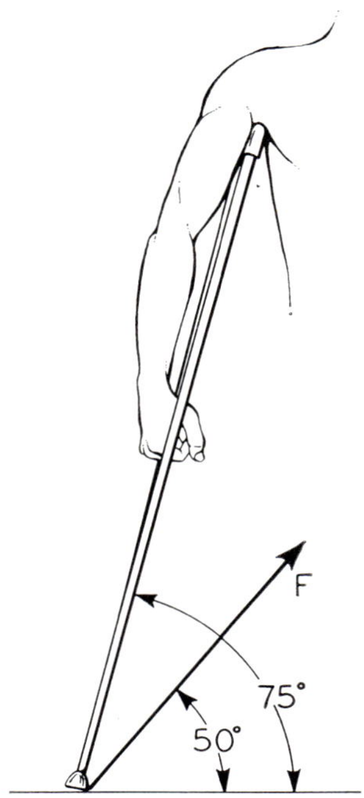

Fig. 1–6. Reactive force against a crutch tip (F).

method. This graphical procedure consists of drawing a line in the vertical direction through the point of application of the crutch force (Fig. 1–7A). A line from the end, or arrow, of the crutch force is then drawn perpendicular to the vertical line. The intersection of the vertical line and the line perpendicular to it from the arrow head defines the magnitude of the vertical component of the crutch force. This component, labeled F_y, is the portion of the crutch force which is effective in supporting the weight of the patient.

If the floor is very smooth, the crutch has a tendency to slip outward, but since the crutch has a rubber tip which increases friction, the floor is able to exert a force inwardly which prevents the crutch from slipping outward. This inwardly directed force is called the friction force. As in the previous example, the friction force is also a component of the total force F. The friction force may be deter-

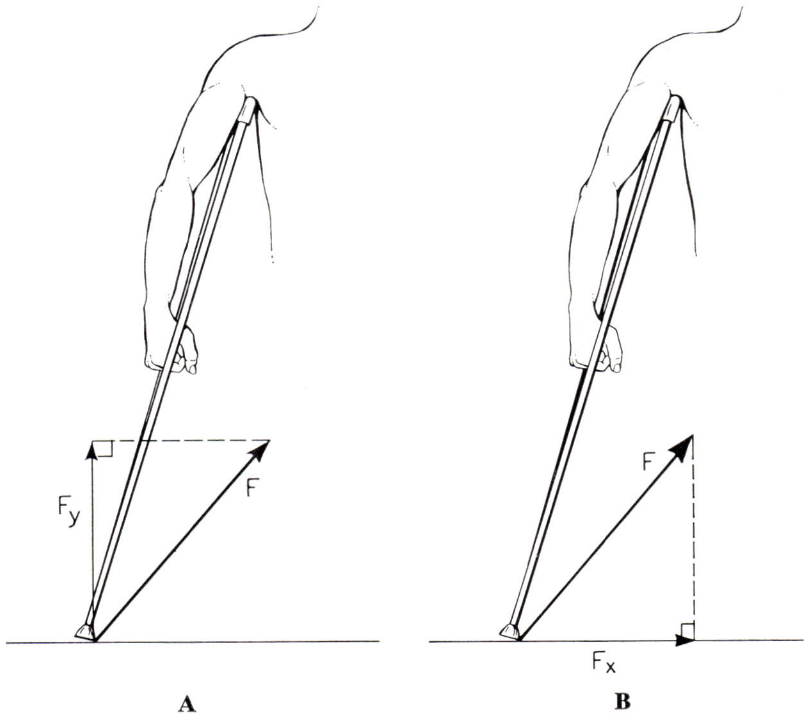

Fig. 1–7. **A,** Vertical (F_y) and **B,** horizontal (F_x) components of force (F).

mined in a manner similar to that used to obtain F_y. A line is drawn through the crutch tip parallel to the floor, that is, in the horizontal direction (Fig. 1–7B). A perpendicular is then drawn from the head of the arrow to this horizontal line. The quantity F_x is then representative of the frictional force developed between the crutch and the floor.

Force F has components in any arbitrary direction that one may choose. In Figure 1–8 an arbitrary direction labeled Z-Z has been chosen. The component of force F in the Z-Z direction has been constructed and is labeled F_z. If the Z-Z direction is specially chosen to be perpendicular to force F, then the component of force F in this perpendicular direction will be zero.

It is very important to realize the difference between the components of a force and its partial sum. In Figure 1–9 vectors A and B, which represent forces, are added by the parallelogram law to produce the force C. C is said to be the vector sum of A and B. Notice, however, that if a component of vector C, namely C_x, is taken along the direction of vector B, C_x is not equal to B.

A force may have an infinite number of components. One is perfectly free to choose a direction and then to find a component of a

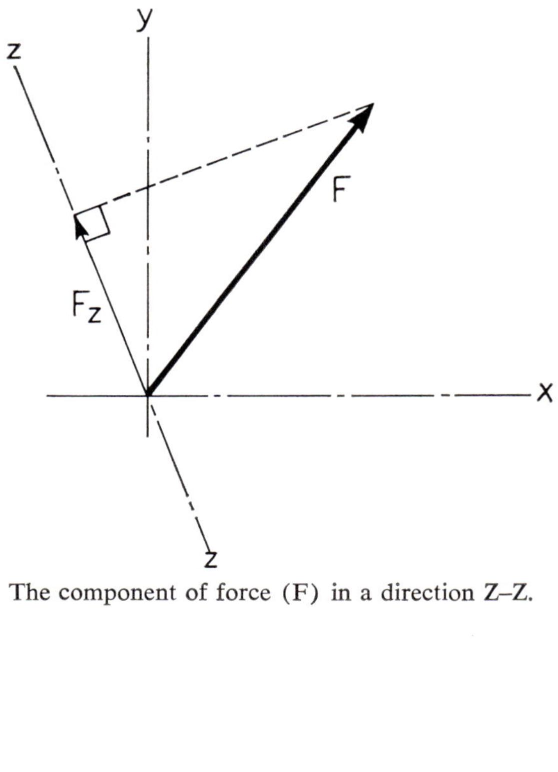

Fig. 1–8. The component of force (F) in a direction Z–Z.

force in that direction. The addition of vectors, however, is not the reverse of the same process.

Components of forces may be calculated algebraically as well as graphically. As an example of the algebraic procedure used to calculate the components of a vector, consider the crutch in Figure 1–10.

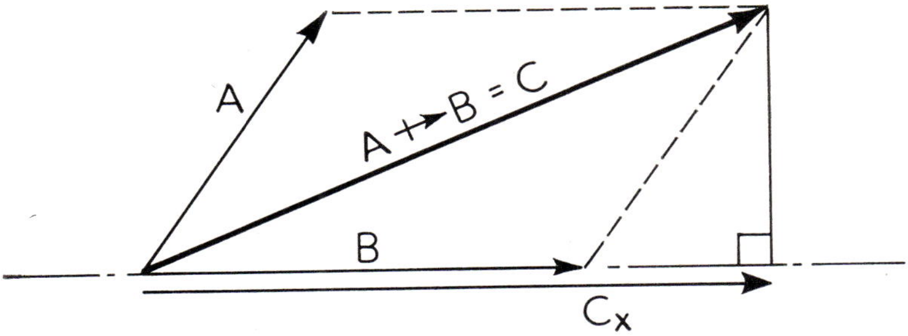

Fig. 1–9. A and B are two vectors whose sum is equal to C. C_x is the component of vector C in the direction of vector B. This component C_x does not equal vector B.

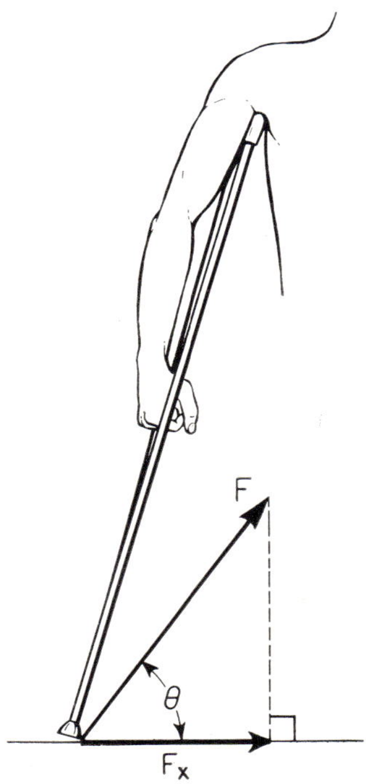

Fig. 1–10. Trigonometric relations may be used to determine components of a force. (F_x) is the component in the horizontal direction; θ is the angle between the force F and the ground.

To find the component f_x, which is the frictional force, the angle θ must either be determined experimentally or measured from the diagram. The ratio of the friction force F_x to the total force F is then equal to the cos θ. This can be written in the algebraic expression

$$\frac{F_x}{F} = \cos\theta$$

Since both F and cos θ are known quantities, F_x may be calculated. Thus

$$F_x = F\cos\theta$$

In the study of biomechanics, applied forces may often be considered as concentrated forces acting at a point. It is important to realize that this is an idealized situation which seldom occurs in nature. For example, the force applied to the head of the femur by the acetabulum is a nonuniformly distributed force. As such, it may accurately be described as a nonuniform pressure. If the object of the

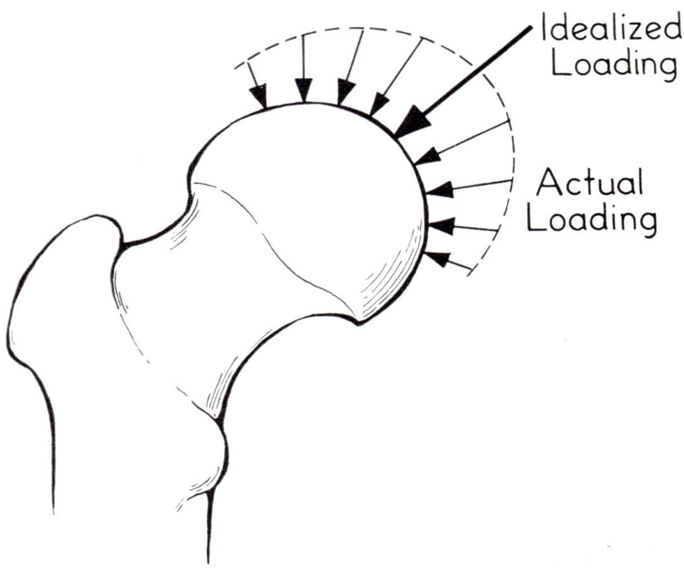

Fig. 1–11. In studying the external effects of forces on bodies, it is sufficient to consider an idealized loading state.

immediate study is not to determine the internal effects in the area of the head and neck of the femur, the actual distributed load can be replaced by an idealized equivalent force vector, as shown in Figure 1–11. The magnitude and direction of the equivalent force vector and its point of application can be determined by integrating the distributed load over its applied area. This idealized single force may be used in the study of the statics and dynamics of the femoral head. It is also reasonable to utilize this idealized loading condition when studying the internal effects within a remote part of the femur. For example, the stress distribution at the knee joint does not depend upon the choice of loading configuration at the head of the femur as long as the two systems are statically equivalent.

In the same manner in which the physical properties of that quantity called force have been described, the properties of a quantity known as a moment or a torque will now be considered. A moment may be visualized as the twisting or turning effect of a force or combination of forces. Moments, like forces, have both internal and external effects upon the body on which they are acting. The external effect of a moment on a body is to change or attempt to change the angular or rotational velocity of the body. The internal effects of

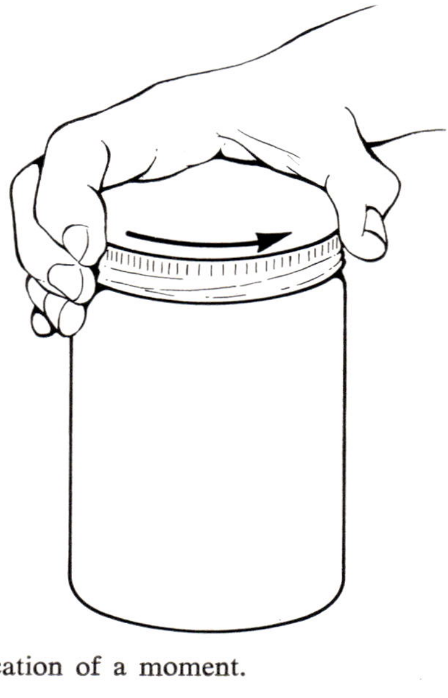

Fig. 1–12. Application of a moment.

moments are to cause a state of strain. Chapter 2 will deal with this aspect.

If a body is rotating at constant angular velocity about a particular axis, or, as a special case, has zero angular velocity, a moment must be applied to that body in order to change its angular velocity. A moment is therefore necessary to produce angular acceleration. Moments possess the four vector characteristics. Magnitude, the first of these characteristics, is somewhat different in its unit designation from the unit designation of the magnitude of a force. The magnitude of a moment is represented by the product of the units of force and distance. Pound-feet or kilogram-centimeters are appropriate units for the magnitude of a moment. By convention, the unit of weight is placed first and the unit of distance is placed second. If the lid of a jar is unscrewed, a moment of 6 pound-inches must be applied to the lid, as shown in Figure 1–12. This is the same as ½ pound-foot since there are 12 pound-inches to a pound-foot. Similarly, there are 10 gram-millimeters in a gram-centimeter and 1,000 gram centimeters in a kilogram-centimeter.

The line of application of a moment is defined by convention (Fig.

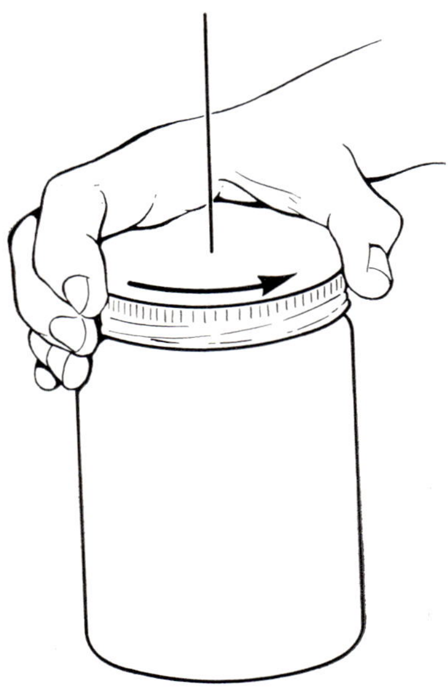

Fig. 1–13. Conventional representation of the line of application.

1–13). In the case of the moment applied to the jar lid, the line of application of the moment is a line perpendicular to the surface of the lid. Although the moment is usually thought of as being applied in the plane of the lid, the line of application of the moment is perpendicular to the plane of the lid. It is easily seen that there are two possible moments which have the same direction; namely, in the case of the cap, unscrewing the cap or tightening the cap. To be able to describe uniquely these two moments, an arbitrary definition for the sense of the moment has been chosen. This definition is based upon the *right-hand rule,* as illustrated in Figure 1–14. If the fingers of the right hand are wrapped in the direction of the application of the moment, *i.e., counterclockwise,* the extended right thumb indicates the sense of the moment vector. The point of application of the moment

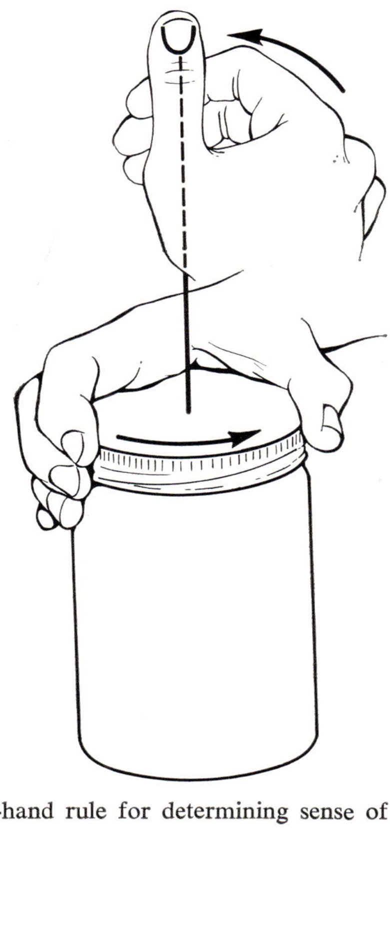

Fig. 1–14. Right-hand rule for determining sense of a moment vector.

vector is taken simply as the intersection of the moment vector and the plane in which the moment is acting.

One effect of a force acting on a body is to create a moment about any other point in the body (Fig. 1–15). The moment produced by a force about a point is equal to the product of the force and the perpendicular distance between the line of application of the force and the point. When a nail plate is loaded at its tip, a moment is produced at the nail-plate junction, as can be seen in Figure 1–15. This moment can be calculated by knowing the force acting on the tip of the nail-plate junction. The moment is the product of the force and the perpendicular distance.

If a 12-kg weight is held in the hand, a moment due to this weight

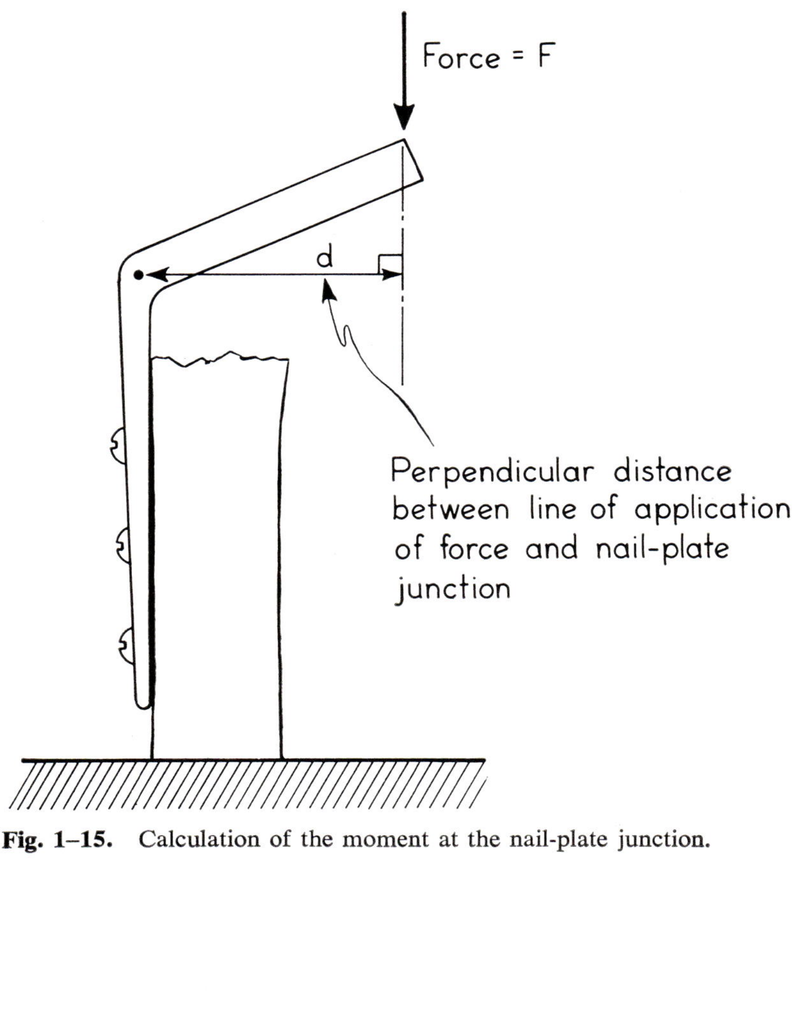

Fig. 1–15. Calculation of the moment at the nail-plate junction.

of 3,000 kg-mm will be developed at a distance of 250 mm at the bicipital tubercle. That is, a weight of 12 kg and a distance of 250 mm produce a moment of 12 times 250 or 3,000 kg-mm. A moment always involves a force and a distance. If a force acts on one part of a body, a moment will be developed at any distance from the point of application of the force in the body.

The solution of biomechanical problems involves the determination of unknown reactive forces and moments. For the purpose of analysis it is always convenient and usually necessary to consider

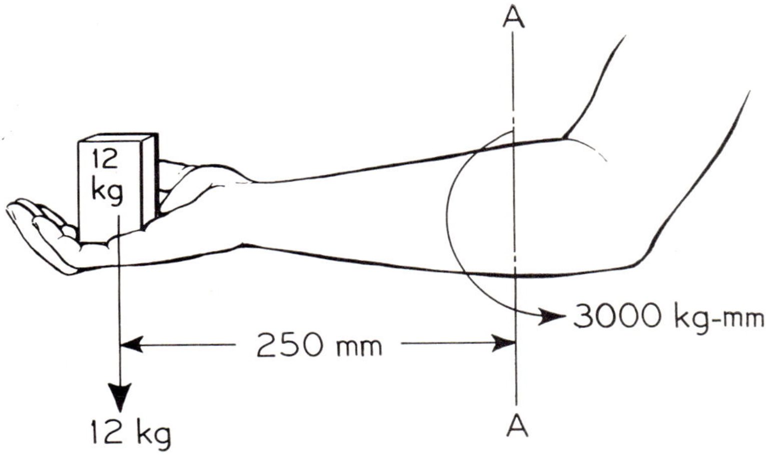

Fig. 1–16. The moment produced at the bicipital tubercle by a 12-kilogram weight held in the hand.

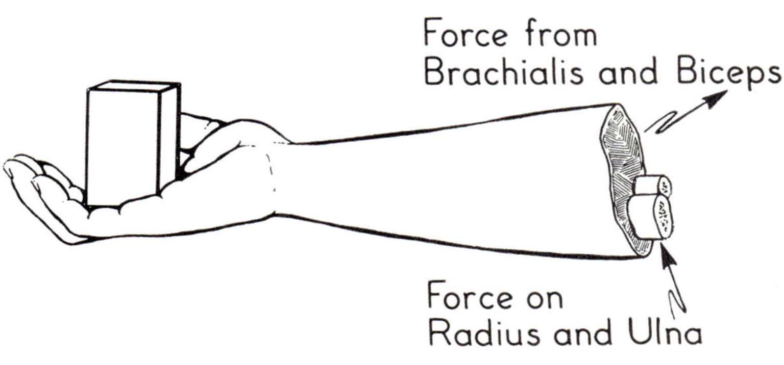

Fig. 1–17. A free body of the forearm.

portions of the system as distinct from the entire system. A portion of the system so considered is called a free body. For example, to analyze forces acting upon the forearm at section A in Figure 1–16, the free body shown in Figure 1–17 is used. In drawing the free body diagram, a seemingly arbitrary geometric boundary is chosen. Actually, any boundary can be chosen if it will aid in the solution; however, it is quite important that the appropriate interacting forces be placed across the chosen boundary. Only forces that cross the boundary of the free body must be considered.

The free body diagram concept is the basic tool in solving all problems in statics. Later its application to dynamic problems will also be shown.

The basic physical principles used in solving problems in statics are:

1. If a body is in *equilibrium,* the *vector sum* of all forces acting on that body is zero. (If a body is in equilibrium the sum of the *components* of all forces in *any direction* is zero.)

$$\Sigma F = 0$$

2. If a body is in equilibrium, the sum of the moments *about any point* on the body is zero.

$$\Sigma M = 0$$

To illustrate the above principles consider the following problem: What are the reactions on the radius and ulna in the proximal forearm in Figure 1–17?

Since the reaction in this region, labeled section A, is unknown, the most general form must be assumed, that is, that a moment M and a force F may exist (Fig. 1–18). The force F is taken to be in some general unknown direction and of unknown magnitude. The moment M is taken to be unknown in magnitude and sense. All forces applied across the boundary of the free body are included.

To solve the problem, F is resolved into two convenient component vectors, F_h and F_v, whose sum is, in this special case, F. When a vector is resolved into two perpendicular components, the sum of the two components equals the original vector. The first principle of equilibrium, that is, the sum of the components of all forces acting on a body in equilibrium is equal to zero, may be applied. The sum of the horizontal components of all forces is zero.

$$\Sigma \text{forces}_{\text{horizontal}} = 0 \tag{1}$$

The sum of the vertical components of all forces is zero.

$$\Sigma \text{forces}_{\text{vertical}} = 0 \tag{2}$$

The second principle of equilibrium will also be used.

$$\Sigma \text{moments} = 0 \tag{3}$$

at any section.

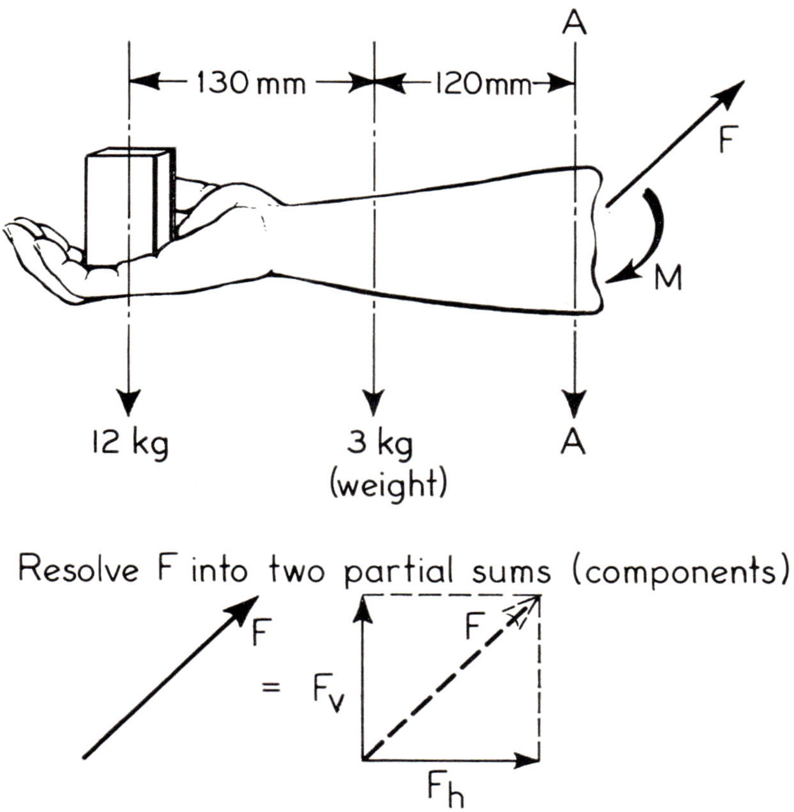

Fig. 1–18. Unknown force (F) and moment (M) acting at section (A) of the free body.

From the first equation

$$F_h = 0 \tag{4}$$

From the second equation

$$-F_v + 12 \text{ kg} + 3 \text{ kg} = 0 \tag{5}$$

(Note: The sign of F_v is negative since its sense is upward.)
Therefore

$$F_v = +15 \text{ kg} \tag{6}$$

indicating that the correct sense for F_v is upward.

Considering moments about an axis through section A, from the third equation

$$(12 \text{ kg})\ (250 \text{ mm}) + (3 \text{ kg})\ (120 \text{ mm}) + (F \text{ kg})\ (0 \text{ mm}) - M = 0 \tag{7}$$

(Note: The sign of M is negative because it is shown to act clockwise, while other moments due to the forces are counterclockwise.)

Thus

$$M = 3360 \text{ kg-mm}$$

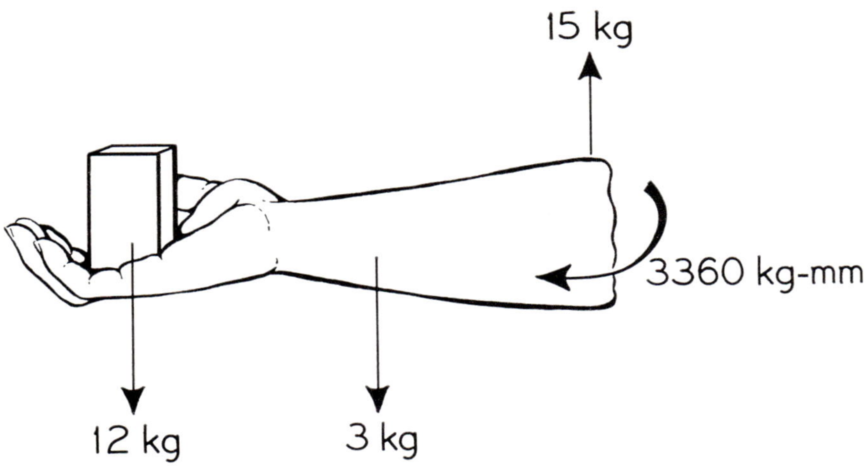

Fig. 1–19. A free body diagram of the forearm with all forces and moments necessary for equilibrium.

The completed free body diagram appears as indicated in Figure 1–19. In answer to the original question, there is a force of 15 kg and a moment of 3,360 kg-mm acting on the proximal forearm when a 12-kg weight is held in the hand.

If a body is in equilibrium under the action of three coplanar forces, they are either concurrent or parallel.

In the sketch of the skull used to determine the forces acting on the temporomandibular joint during normal chewing, forces are exerted by the temporalis and masseter muscles (Fig. 1–20). A force is also applied to the teeth. These forces are in equilibrium and,

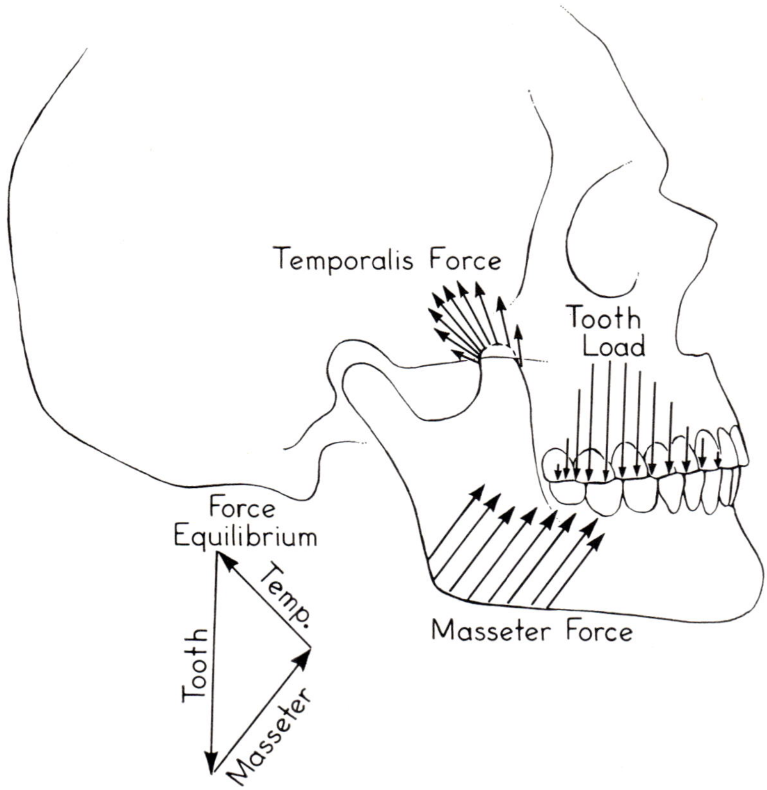

Fig. 1–20. Forces exerted in chewing. When the idealized forces are placed head-to-tail, they form a closed triangle indicating a condition of equilibrium.

therefore, equal zero. This can be seen by placing the forces head to tail in their proper orientation and noting that they form a closed triangle. If the triangle did not close, the sum of the forces would not be equal to zero. With this simple technique any one of the three forces can be calculated if the other two are given. Therefore, in normal chewing it is not necessary to have a force acting on the temporomandibular joint to achieve equilibrium.

When cervical traction is applied to a chin strap, equilibrium is achieved by a reactive force created at the temporomandibular joint (Fig. 1–21). The reactive forces are in equilibrium, as demonstrated

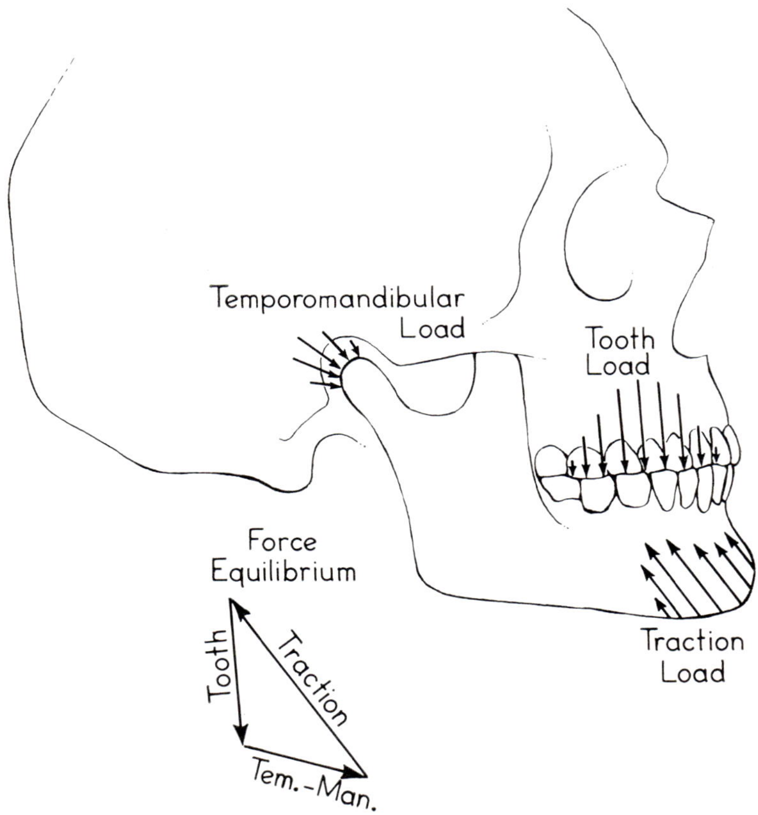

Fig. 1–21. The force system in cervical traction. The equilibrium state is achieved only through the application of a force acting on the temporomandibular joint.

by the closed triangle, and the force acting on the temporomandibular joint can be calculated since the other two forces are known. A prolonged load on the temporomandibular joint produces pain. To eliminate the reactive force at the temporomandibular joint, another equilibrium condition can be achieved in cervical traction by interposing a bite plate between the teeth (Fig. 1–22). The bite plate produces an alteration of the direction of the load on the teeth which allows this force to be equal and opposite to the traction force without a load on the temporomandibular joint.

Another useful clinical application of the concepts of free body analysis is that of calculating the load borne by the femoral head. It is possible to construct a free body diagram of the hip joint to calculate the unknown forces and moments acting on the head of the

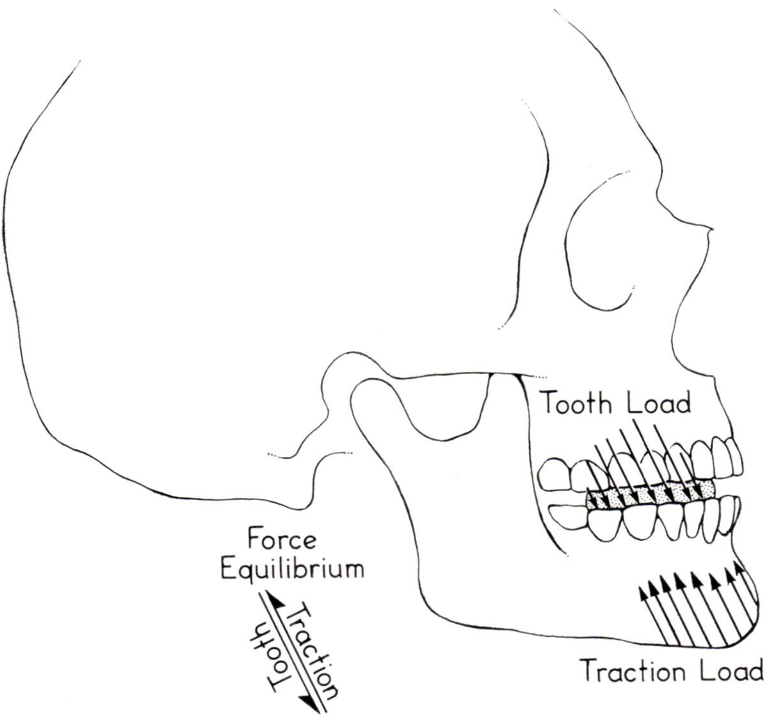

Fig. 1–22. The effect of a bite plate during cervical traction. Equilibrium is achieved through the mechanism of a bite plate, which allows the tooth force to produce a force equal but opposite to the traction force.

femur during different conditions of equilibrium. Consider the problem of finding the force on the femoral head during the single stance phase of gait. It is known that a lower extremity weighs approximately $\frac{1}{6}$ of the total body weight (Fig. 1–23). It is also obvious that the reactive force W of the floor against the foot is equal to the body weight. If the entire system shown in Figure 1–23 is considered a free body, the desired force does not appear as an external force. Therefore, the system must be divided into two free bodies. To create a free body diagram, not only must the hip joint be separated, but adjacent muscular and ligamentous structures must also be divided. The free body boundary extends through a portion of the hip joint as well as through the pelvitrochanteric muscles (Fig. 1–24). Since it is known that there is a reactive force at the hip joint, it must also be concluded that a reactive force is present in the other structures adjacent to the hip joint. The following information is available to construct the free body diagram shown in Figure 1–25.

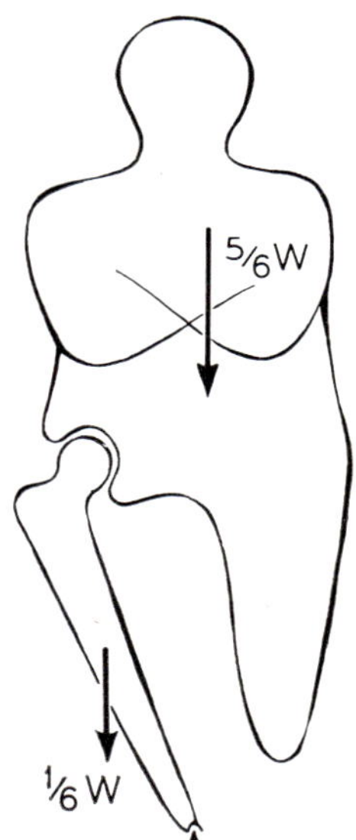

Fig. 1–23. The reactive internal force—the force on the head of the femur—does not appear. W is the body weight.

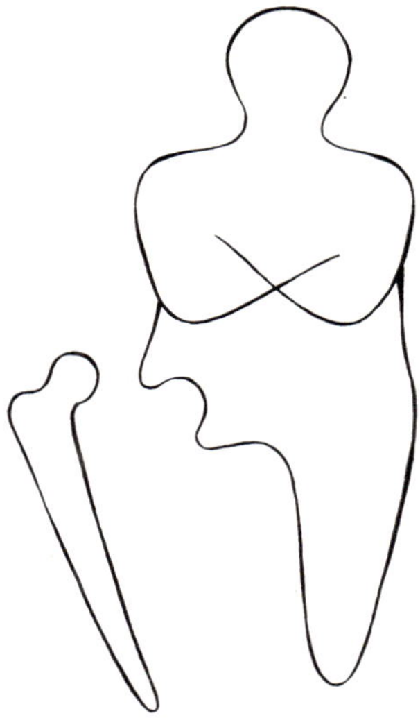

Fig. 1–24. Division of the single free body seen in Figure 1–23 into two free bodies so that a boundary passes through the line of application of the desired force.

On the free body of the extremity there is a reactive force F at the hip joint of known magnitude and direction.

There is a muscle force P with unknown magnitude, but with known direction, since the force developed by a muscle is known to act along the long axis of the muscle.

There is a reactive force W of the floor against the foot with known magnitude and direction.

There is a gravitational force of $\frac{1}{6}$ of W acting on the limb with known magnitude and direction.

There are only three forces acting on the free body of the trunk illustrated in Figure 1–26. They are: the remainder of the body weight, $\frac{5}{6}$ of W; the reactive force at the hip joint F; and the reactive force P of the abductor muscles.

The problem is to solve for the unknown components, namely the magnitude and direction of the reactive force against the hip joint F and the magnitude of the muscle force P. The muscle force P can be found by considering the moment equation of equilibrium as applied to the free body of the trunk. Because of equilibrium, the sum of the

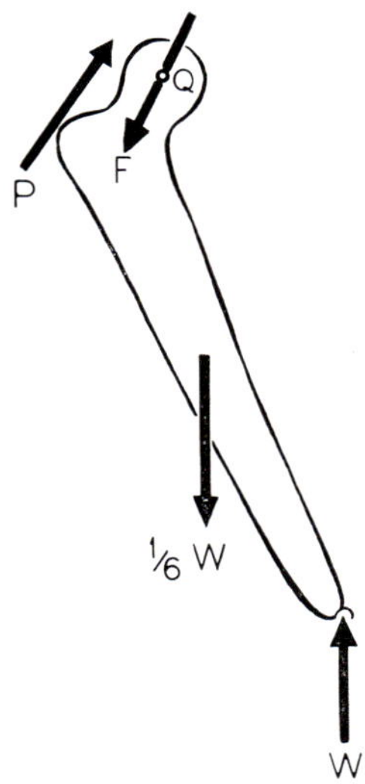

Fig. 1–25. Free body of lower extremity which includes all forces passing through the boundary of the body. P is the abductor muscle force with known direction. One sixth of W is the weight of the leg. W is body weight of the assumed geometric center. F is the reactive force assumed to pass through Q, the center of the joint with unknown magnitude and direction.

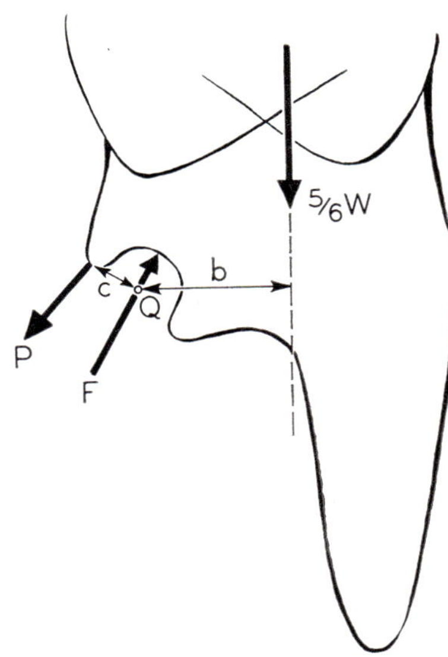

Fig. 1–26. A free body of the remainder of the body. Q, P, and W are defined in Figure 1–25, as is F. The perpendicular distance between the line of action of the muscle force P and the reactive force passing through Q is c. The perpendicular distance between the line of action of the gravitational force of $5/6$ of W and point Q is b.

moments about point 0 is equal to zero, or force P times the distance, which is represented as c, minus a force $\frac{5}{6}$ of the body weight times the distance b is equal to zero.

$$\Sigma M = 0$$

$$Pc - \frac{5}{6} Wb = 0$$

Solving for P,

$$P = \frac{5}{6} W \frac{b}{c}$$

From anatomical and x-ray studies, the usual range of values for the ratio of b over c is between 2 and 3.5; thus, the value of P will be between 1.7 W and 2.9 W. Note that this muscle force is more than 2 times body weight. This is necessary to balance the body over the hip joint.

The free body of the limb can be considered now, and the reactive force acting against the head of the femur can be determined (Fig. 1–27). To solve for the force F, the reactive force against the femoral

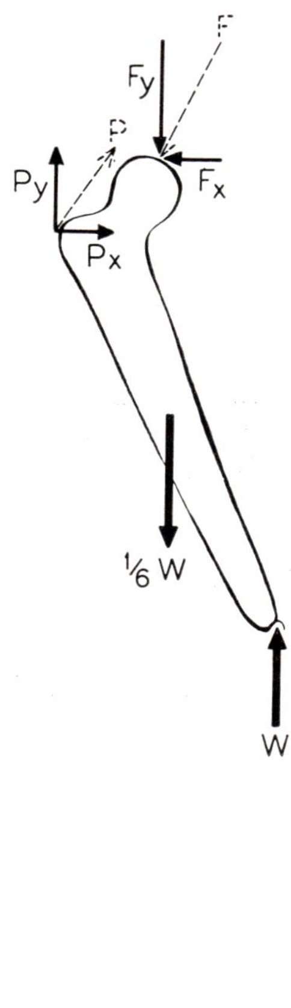

Fig. 1–27. A free body of the lower extremity. The muscle force P and the reactive force F are separated into their vertical (y) and horizontal (x) components.

head, it is necessary to utilize equilibrium conditions; that is, the sum of the forces in the horizontal, or X, direction is equal to zero, and the sum of the forces in the vertical, or Y, direction is equal to zero. Two components of the force F, the horizontal and vertical components, will be determined. The total force F acting against the femoral head can then be determined by vector addition. It should be noted that the magnitude of P_x and P_y, that is, components of the muscle force P in the horizontal and vertical directions, depend on the original direction of the muscle force. For this particular example, the value of 30 degrees has been assumed for the angle which the muscle makes with the vertical.

P_x, the component in the horizontal direction, is related by

$$P_x = P \sin 30° \tag{8}$$

and is therefore equal to .50P.

P_y is found by utilizing the cosine function

$$P_y = P \cos 30° \tag{9}$$

and is equal to .86P. From the previous calculations, P is found to vary from between 1.7 W and 2.9 W. For purposes of this calculation, it shall be assumed that

$$P = 2W \tag{10}$$

That is, the muscle force is equal to 2 times the body weight. Therefore

$$P_x = .5P = W \tag{11}$$

and

$$P_y = .867P = 1.73W \tag{12}$$

Now by considering the equilibrium equation, the sum of the forces in the X direction is equal to zero, the following equation can be formed.

$$P_x - F_x = 0 \tag{13}$$

Therefore

$$F_x = P_x$$

and from equation (11)

$$F_x = W \tag{14}$$

Now consider equilibrium in the vertical, or Y, direction. The following equation can be formed by setting the sum of the vertical force components equal to zero.

$$P_y - F_y - \tfrac{1}{6}\,W + W = 0 \qquad (15)$$

From equation (12)

$$F_y = 1.73W$$

Using this value in equation (15), F_y may be found to be

$$F_y = 2.56W$$

Now using the Pythagorean theorem

$$F^2 = F_x^2 + F_y^2$$

or

$$F = 2.75W \qquad (16)$$

Thus, the reactive force against the head of the femur is equal to 2.75 times the body weight. It should be noted that the only changes in geometry that can affect F are changes that alter the direction of P or its magnitude, that is, the ratio b over c.

To determine the direction of the resultant acting against the head of the femur, it is necessary to utilize the tangent function as seen in Figure 1–28. In this case, the angle designated as angle α between the line of force acting on the femoral head and the horizontal has its tangent given by the ratio of F_y over F_x, that is, the opposite over the adjacent sides (Fig. 1–27). This ratio is equal to 2.57, which is the tangent of 69 degrees.

On the basis of the above calculations, during the single stance phase of gait the idealized equivalent force on the femoral head is 2.75 times the body weight and is applied as an angle of 69 degrees from the horizontal.

As previously stated (page 19), if only three nonparallel forces act on a body they must be concurrent. This can be proved by examining Figure 1–29. If this body is to be in equilibrium, the sum of the moments about any point must equal zero. However, the sum of the

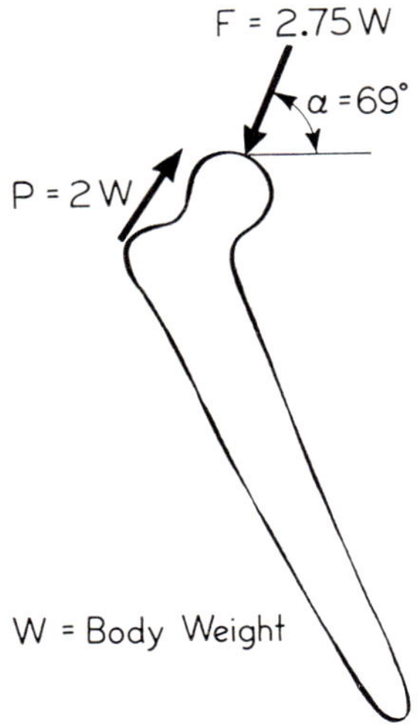

Fig. 1–28. Free body of lower limb with the resulting muscle force and reactive force against the head of the femur, as shown.

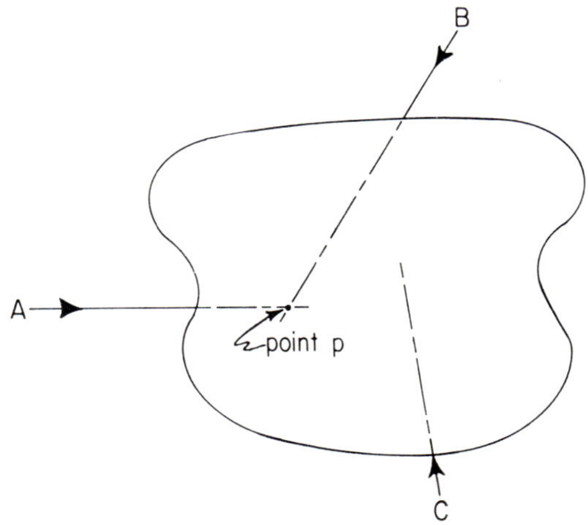

Fig. 1–29. Free body with three nonconcurrent forces.

moments about point p cannot be zero since force C produces a moment about point p, while forces A and B do not. Therefore, the body as shown cannot be in equilibrium unless all three forces pass through point p. This fact can be used in the solution of problems involving only three forces.

In Figure 1–30 a man is shown climbing stairs. It is desirable to know the contact force between the joint surfaces and the force on the patella tendon. A free body analysis of the lower leg (Fig. 1–31) is performed but the weight of the lower limb portion (approximately $1/15$ of body weight) is ignored. This reduces the analysis to a problem involving three forces. (Lower limb weight may be ignored without loss of accuracy since it is an order of magnitude smaller than any other force.) The three forces acting on the free body are:

W—Ground reaction equal to body weight, known magnitude, line of application, sense, and point of application.

P—Tension in the patella tendon. Known line of application, sense, and point of application. Unknown magnitude.

J—Joint reaction force. Known point of application. Unknown line of application, sense, and magnitude.

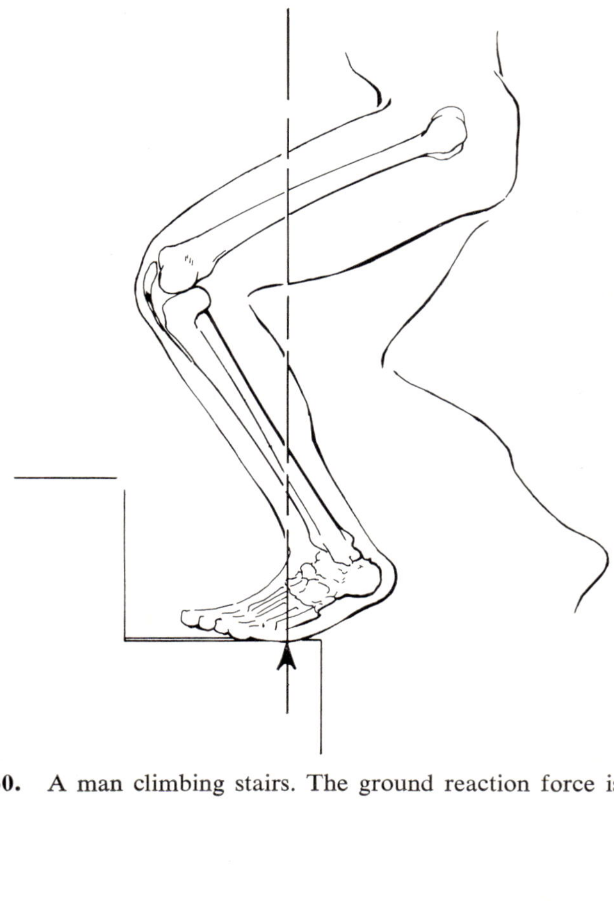

Fig. 1–30. A man climbing stairs. The ground reaction force is shown.

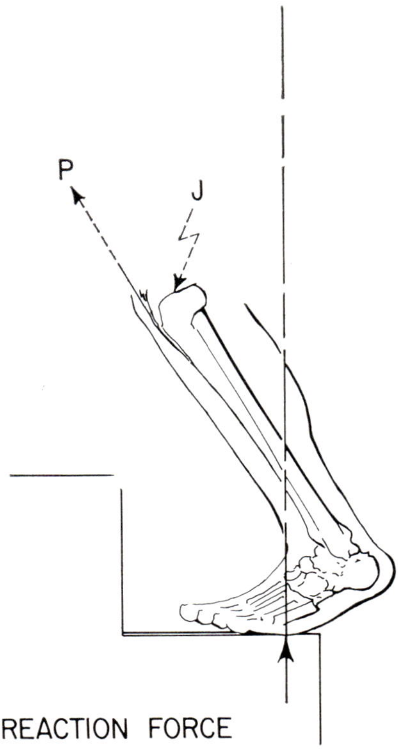

Fig. 1–31. Free body of the lower limb of a man climbing stairs. The direction of the patella tendon force is known, but the direction of the joint reaction is not.

The line of application of force J may be determined by first obtaining the intersection point (p) of forces W and P (Fig. 1–32). The line of application of force J must pass through both its point of application on the joint surface and point p (Fig. 1–33). With the line of application of force J determined, it is now possible to construct a triangle of forces.

First force W is drawn (Fig. 1–34) with the proper line of application and sense. Next, a vector representing force P is added to W. The line of application and sense of P are known but the vector cannot be terminated since the magnitude is unknown. It is known, however, that when force vector J is added to the sum of W and P, the

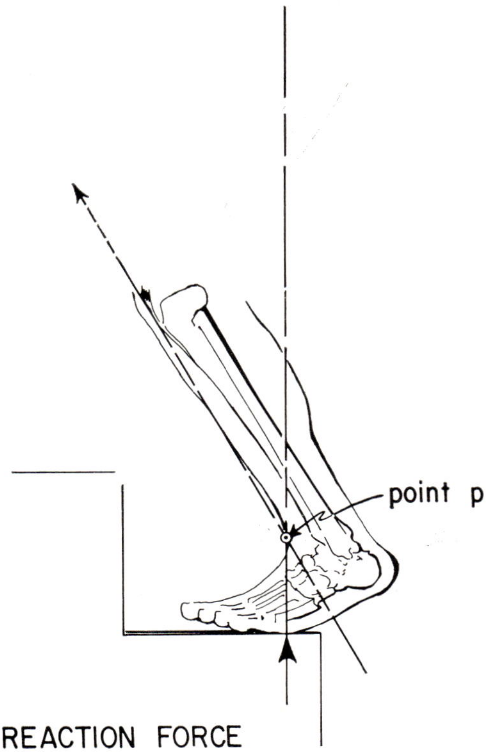

Fig. 1–32. Extending the lines of application of the patella tendon force and the reaction force locates their point of intersection (p).

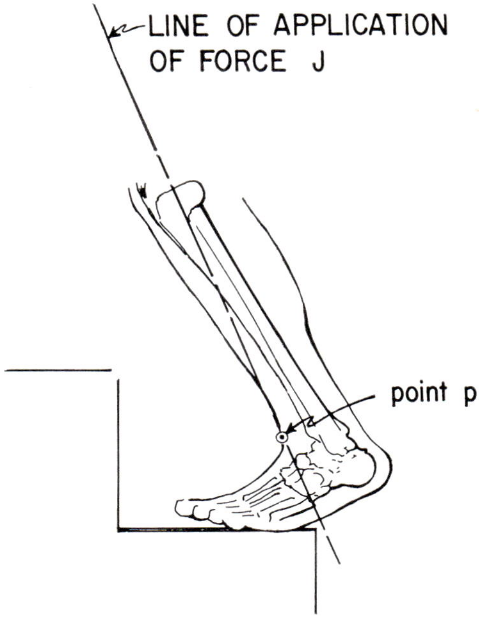

Fig. 1–33. The line of application of the joint reaction force (J) is obtained by connecting point p with the contact point between the tibia and the femur.

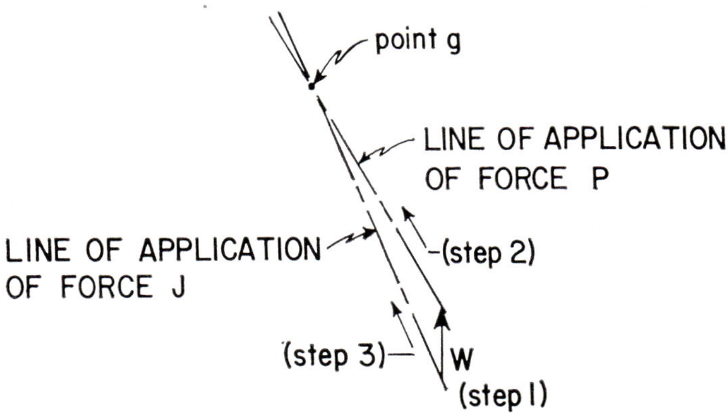

Fig. 1–34. The line of application of force P is drawn at the head of force W, while the line of application of force J is drawn at the base of force W. The intersection of these two lines (point g) defines the length of these force vectors.

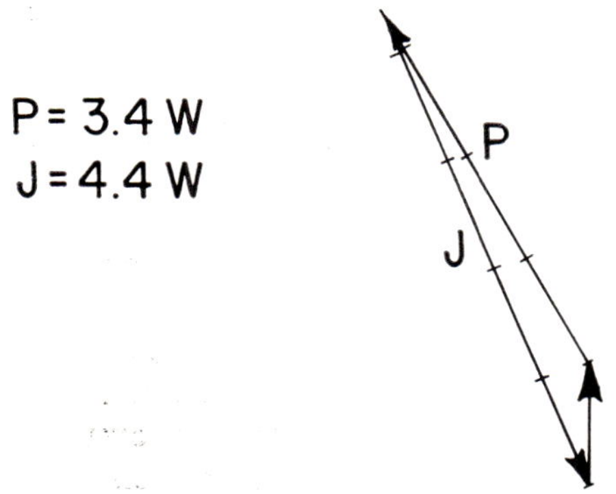

Fig. 1–35. The force triangle may now be scaled to obtain the values of P and J.

triangle must close. (The head of vector P must touch the origin of vector W.) Therefore, starting at the origin of W a line is drawn parallel to the line of application of force J. This line intersects vector P at point q. Point q must be the head of vector P and the origin of vector J (Fig. 1–35). The magnitudes of forces P and J can now be scaled from the drawing.

Dynamics is the study of the action of forces on bodies not in equilibrium. A body not in equilibrium is accelerating. An extension of Newton's second law of motion is that the time rate of change of velocity, that is, the acceleration, is proportional to the resultant applied force. This is written as the well-known expression:

$$F = M \cdot a$$

in which F is a force measured in kilograms (kg), M is the mass measured in kilograms per second squared divided by millimeters ($kg\text{-}sec^2/mm$), and a is the acceleration measured in millimeters per second squared (mm/sec^2). The physical interpretation of this mathematical statement is that whenever a body suffers a change of velocity, that is, a change in either the magnitude or direction of the velocity, there must be a resultant force acting on the body. It must be remembered that velocity is a vector quantity and as such possesses magnitude, direction, and point of application. A change in either magnitude (speed) or direction constitutes a change in velocity. When a person moves his leg through one cycle of gait, muscles must apply forces both to accelerate and to decelerate the leg. When solving problems involving linear acceleration, it is usually convenient to consider the entire mass of the body as concentrated at a point called the body's center of gravity. The body may then be considered a point mass for analytical purposes.

When analyzing the dynamics of gait, it is convenient to utilize this concept of point mass and to discuss motion of the center of gravity of the body in the analysis. The terms mass and body weight are not synonymous. Body weight is a force and is equal to the body mass multiplied by the gravitational acceleration of the earth. A man who weighs 150 pounds is being attracted towards the center of the earth with a 150-pound force. If he stands on a scale, the surface of the scale pushes up on his feet with 150 pounds. His mass can be calculated by knowing that if the scale reaction were absent, he would accelerate downward at 32.2 feet per second (gravitational acceleration). Therefore

$$\underset{\text{(Force)}}{150 \text{ lbs}} = \text{man's mass} \times \underset{\text{(Acceleration)}}{32.2 \frac{\text{ft}}{\text{sec}^2}}$$

or

$$\text{Man's mass} = 4.66 \frac{\text{lb} \cdot \text{sec}^2}{\text{ft}}$$

The combination of units $\frac{\text{lb} \cdot \text{sec}^2}{\text{ft}}$ is called a slug.

A further extension of Newton's second law used to describe rotating systems is

$$T = I \cdot \alpha \tag{17}$$

where T is the torque or moment, I is the mass moment of inertia and α is angular acceleration. T, torque, has units such as kg-mm or ft-lbs. I, the mass moment of inertia, has units such as kg-mm-sec^2 or lb-ft-sec^2. Alpha, the angular acceleration, has units of $\left(\frac{I}{\text{sec}^2}\right)$.

The mass moment of inertia is calculated as the integral of the mass of each particle of the body multiplied by the square of its distance to the axis of rotation. This means that the ability of a body to resist changes in its angular velocity, *i.e.,* undergo angular acceleration, is dependent not only upon the amount of mass the body possesses, but also upon the distribution of the mass. The greater the concentration of mass at the periphery, the greater the mass moment of inertia and the more difficult it is to change the angular velocity. A tightrope walker achieves stability, that is, minimal change in angular velocity about the tightrope, by holding a long horizontal pole. This gives him a much greater mass moment of inertia about the tightrope. A fly wheel is made with a massive rim, giving it a higher mass moment of inertia for its given mass.

Newton's second law in the form of equation (17) states simply that if a body is to undergo a change in its angular velocity, *i.e.,* angular acceleration, there must be resultant torques acting on the body (Fig. 1–36). If the positions of the limb in Figure 1–36 are noted, they can be seen to exhibit an acceleration and deceleration of the leg during gait. The fact that the leg undergoes angular acceleration is demonstrated by noting that the angle which the leg makes with the ground has a cyclic variation. After the mass moment of inertia of the leg and the angular acceleration is obtained graphically

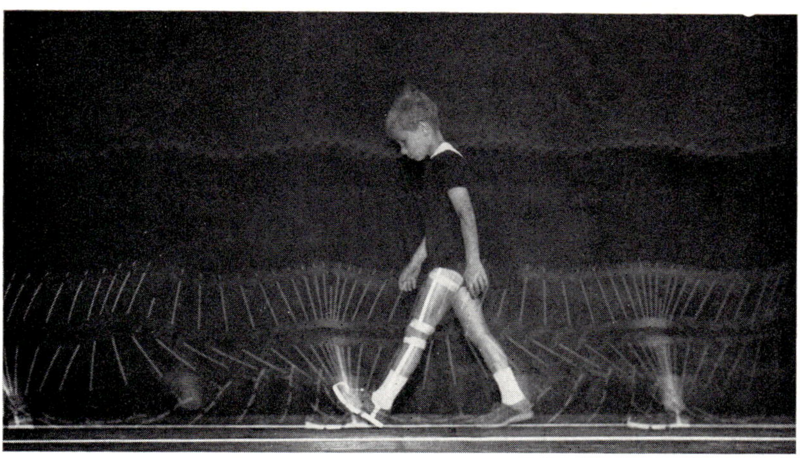

Fig. 1–36. Gait study, utilizing strobe photography, of a subject walking. Reflective targets are placed on the various body segments. Linear and angular displacements may be measured from successive lines and points.

from this diagram, the torque needed at the knee joint to carry the leg through the cycle of gait can be calculated. This value is of interest in designing external and internal knee joint prostheses and transplanting tendons to replace lost quadriceps function.

The principles of dynamics make possible the calculation of the muscular forces necessary to produce such varying body functions as walking, lifting, running, or any similar kinematic task. As an example, the torque at the knee joint necessary to produce normal motion of the tibia during the swing phase of gait will be examined. By use of time-lapse photography, determination of the gross relative positions of the leg during the gait cycle as illustrated in Figure 1–36 is possible.

If these positions are plotted, a graph of the angular position of the lower leg, such as the one in Figure 1–37, may be obtained. By measuring the slope of the displacement curve at each point of the graph, the angular velocity of the foot may be established. The slope of the displacement time curve represents the change in angular position divided by the time interval over which the change occurs. This is the angular velocity. If the graph of the angular velocity versus time is then plotted, the same procedure may be repeated. The slope of the angular velocity versus time curve may be obtained, and this slope represents the angular acceleration. The lower curve in Figure 1–37 shows that, at time equal to 0.5 seconds, which is just before heel strike, the angular acceleration is —45 radians per second^2. For this subject the mass moment of inertia for the lower leg taken about an axis of rotation through the knee is 2.03×10^3 gr • cm • sec^2, using Equation 17,

$$\text{Torque} = 2.03 \times 10^3 \text{ gr}\cdot\text{cm}\cdot\text{sec}^2 \times (-45 \text{ rad/sec}^2)$$
$$\text{Torque} = 9.1 \times 10^4 \text{ gr}\cdot\text{cm} = 91. \text{ kg}\cdot\text{cm}$$

This is the total torque which is exerted about the knee in this instant of the gait cycle. The forces which produce this torque are the gravitational or weight forces, the tension in the muscles traversing the knee joint, the joint reactions, and the elastic forces in the ligaments surrounding the knee joint.

During strenuous activity large joint reactions may be developed due to muscle forces. To illustrate the procedure of calculating joint reactions due to dynamic forces, the knee joint will be used as an example (Fig. 1–38). The maximum angular acceleration that occurs

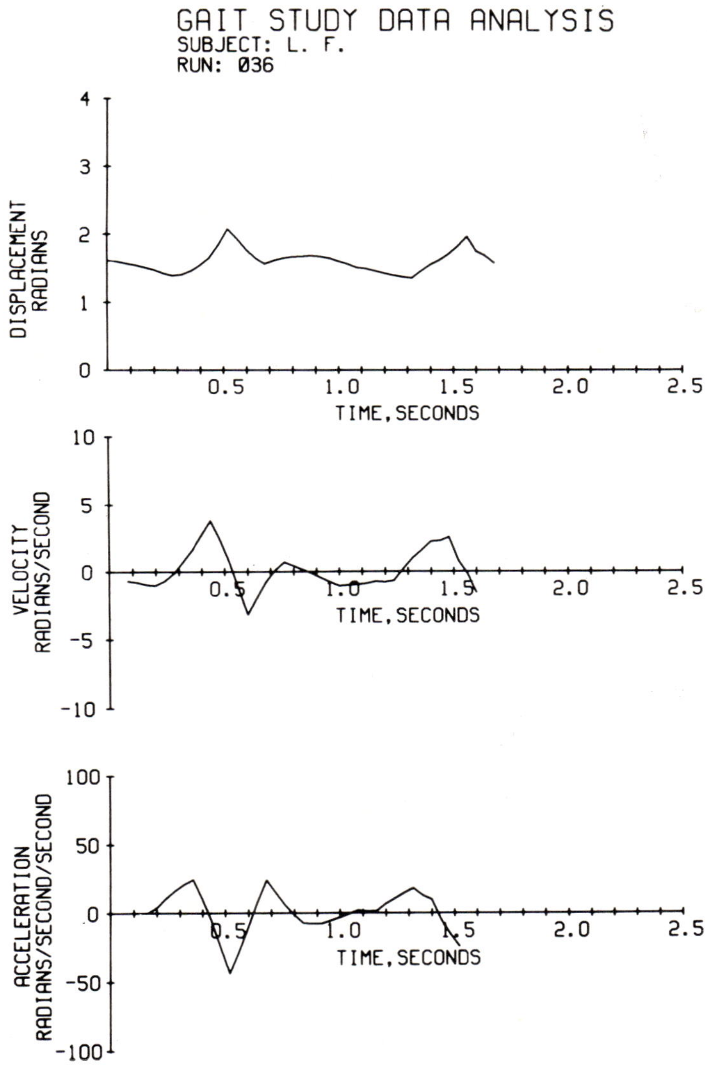

Fig. 1–37. A plot of the angular position, velocity, and acceleration of the tibia during the gait cycle. The reference line (zero displacement) is taken to be a vertical line.

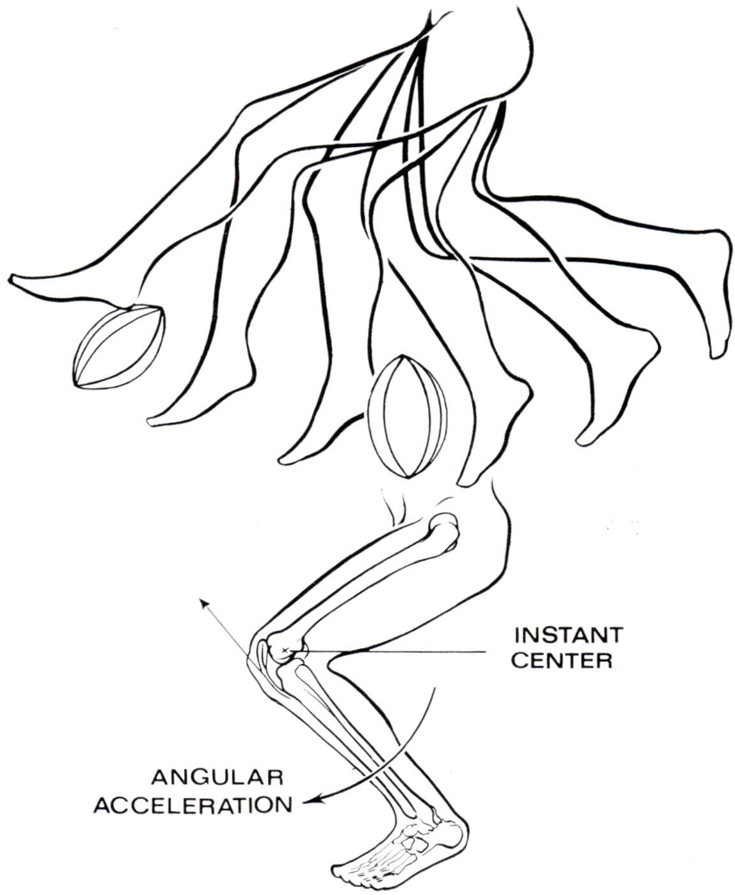

Fig. 1–38. The tension in the patella tendon needed to accelerate the leg during kicking may be calculated if the mass moment of inertia of the leg and its angular accelerations is known.

during the act of kicking a football has been measured at 453 radians per second^2 in an athletic male. This acceleration occurs when the lower leg is almost vertical. The mass moment of inertia about the knee joint center is found to be 3.5×10^3 gr • cm • sec^2 in this subject. The torque necessary to produce this angular acceleration can be calculated using Equation 17.

Thus

$$\text{Torque} = 3.5 \times 10^3 \text{ gr}\cdot\text{cm}\cdot\text{sec}^2 \times 453 \frac{\text{rad}}{\text{sec}^2}$$

or

$$\text{Torque} = 15.8 \times 10^5 \text{ gr}\cdot\text{cm}$$

The result is 1,580 kg-cm.

Since the weight force acts through the center of rotation when the lower leg is vertical, the only force to have a moment about the center of rotation of the knee is the tension force in the patella tendon. Since the perpendicular distance from this force vector to the center of rotation is 5.0 cm in the subject measured, the force on the patella tendon can be found.

$$\text{Force} \times \text{distance} = \text{moment}$$

$$\text{Force} = \frac{\text{Moment}}{\text{Distance}}$$

$$\text{Force} = \frac{1{,}580 \text{ kg-cm}}{5.0 \text{ cm}}$$

$$\text{Force} = 316 \text{ kg}$$

Since the body weight of this subject is approximately 100 kg, the tendon force is over 3 times body weight. A free body analysis of the lower leg shows that, since the weight of the lower limb is 5 kg, and the radial acceleration is almost zero at this instant, the force at the joint interface is 311 kg.

2. Elasticity

The orthopaedist deals with many solid structures. They may be composed of biological materials such as bone and cartilage, metals such as steel, or plastics. The study of such materials is called "solid mechanics," and the aspects dealt with in this book are called strength of materials. It is important to have a precise knowledge of the reactions of these materials to loading conditions encountered in the human body. A force acting on a body has both an external effect and an internal effect. (The external effects were discussed in Chapter 1.) Strength of materials concerns itself with the internal effects of forces acting on bodies.

In biological materials, the internal effect of the externally applied load or force is a change in shape of the body associated with the production of internal forces. It is necessary to define precisely these physical phenomena that occur in solids under loads.

Stress is defined as the internal force per unit area which a part of a body on one side of a plane exerts on that part of the body on the other side of the plane. Note that from this definition, stress is intimately associated with an area or plane. The concept of stress demands a particular plane over which the stress is distributed.

A stress is not a load; it is the result of a load. In order for a material to resist any type of loading, it must develop an internal stress. There are two types of stress: normal stress and shear stress. Normal stress is the force per unit area acting perpendicular to the area under consideration, as shown in Figure 2–1. The units for stress are kilograms per millimeter squared $\left(\frac{\text{kg}}{\text{mm}^2}\right)$ or pounds per square inch (psi). These units are the same as the units of

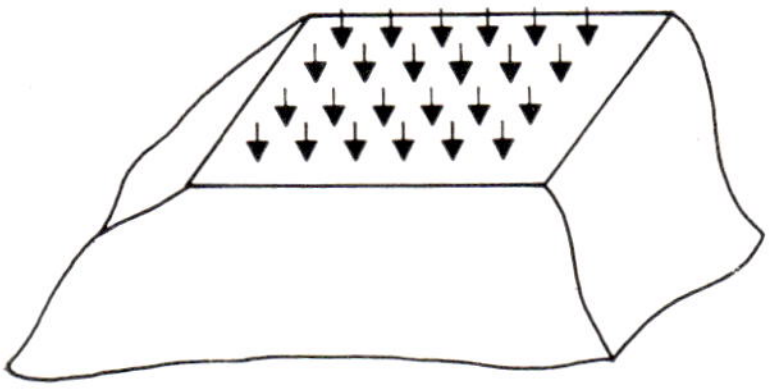

Fig. 2–1. Normal stress. A distributed force acting in a direction perpendicular to a given area.

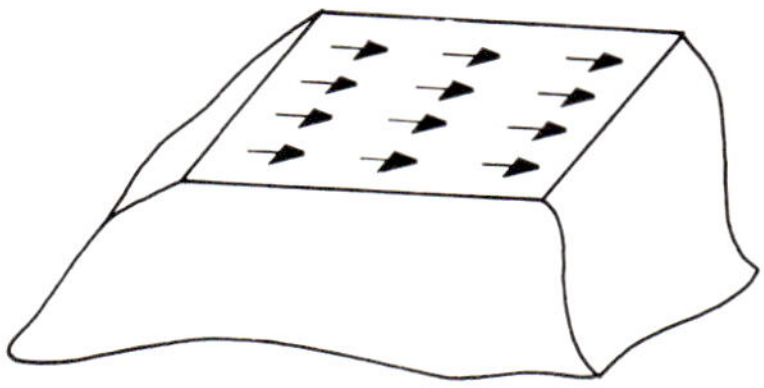

Fig. 2–2. Shear stress. A distributed force acting in a direction parallel to a given area.

pressure. Pressure is a normal stress which acts on the surface of an object. Normal stress is designated by the Greek letter sigma (σ).

The second type of stress is shear stress, which is defined as a force per unit area acting parallel to the area under consideration (Fig. 2–2). Shear stress is designated by the Greek letter Tau (τ). The units of shear stress are the same as the units of normal stress; that is, kilograms per millimeter squared or pounds per square inch. The state of stress on a plane in a body is completely described by the shear stress and normal stress on that plane.

Strain is the change in geometric configuration that a body undergoes. One method of causing strain is the application of external forces. Temperature change also causes strain in a body. Strain can be measured directly. This change can be seen in the skin in the vicinity of a joint as the joint is carried through a range of motion. Markings made on the skin in the region of a joint change size and shape when the joint is moved.

There are two different types of strain. Longitudinal strain is defined as the ratio of the change in length of the line to its original length (Fig. 2–3). Since strain is the change in length, measured in inches or millimeters divided by the original length, also measured in inches or millimeters, the units of strain are inches per inch or millimeters per millimeters. Longitudinal strain is designated by the Greek letter epsilon (ϵ).

Shear strain is defined as the angular deformation suffered by a right angle. That is, if an element of a structure is subjected to a

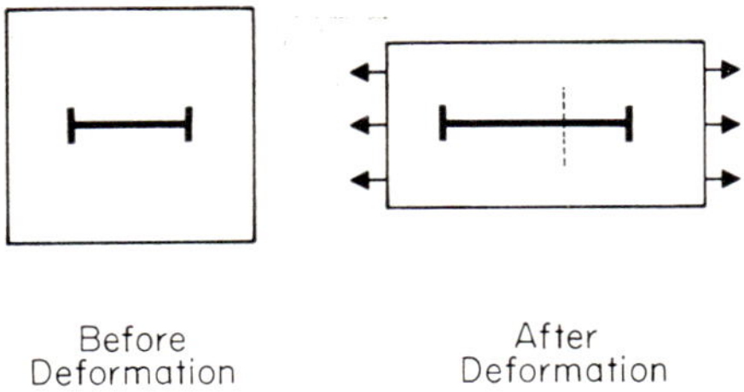

Fig. 2–3. Longitudinal strain. Under the influence of an externally applied force, a body undergoes a change in geometry. One such change is the elongation of straight lines in the body.

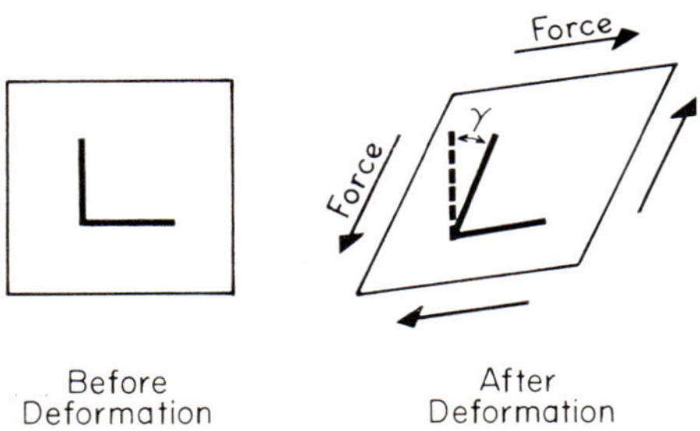

Fig. 2–4. Shear strain. Under the application of externally applied load, lines originally at right angles change their orientation and the angle increases or decreases. Angle γ is the shear strain.

shear strain, two lines of this element which were originally at right angles will now undergo relative angular motion. Shear strain is designated by the Greek letter gamma (γ). This is shown graphically in Figure 2–4. The right angles deform into acute or obtuse angles, depending on their position.

Shear strain is measured as a change in an angle. The units of measurement are radians, not degrees. One radian is defined as the angle whose arc length is equal to the radius of the circle. Radians are the natural mathematical system of angular measurement. The units of "degrees," although a useful and common measurement, cannot easily be used in stress analysis or kinematic analysis.

Strain may be measured by several techniques. Wire strain gauges are fine filaments mounted on a supporting structure. The entire strain gauge is glued to the surface upon which the strain is to be measured. As the underlying surface is strained, the length of adherent wire is changed. This change in length of the wire produces a change in its electrical resistance which may be detected through the use of a Wheatstone bridge (Fig. 2–5).

Another method of measuring strain is by the use of a brittle lacquer coating which cracks at a certain threshold elongation. The

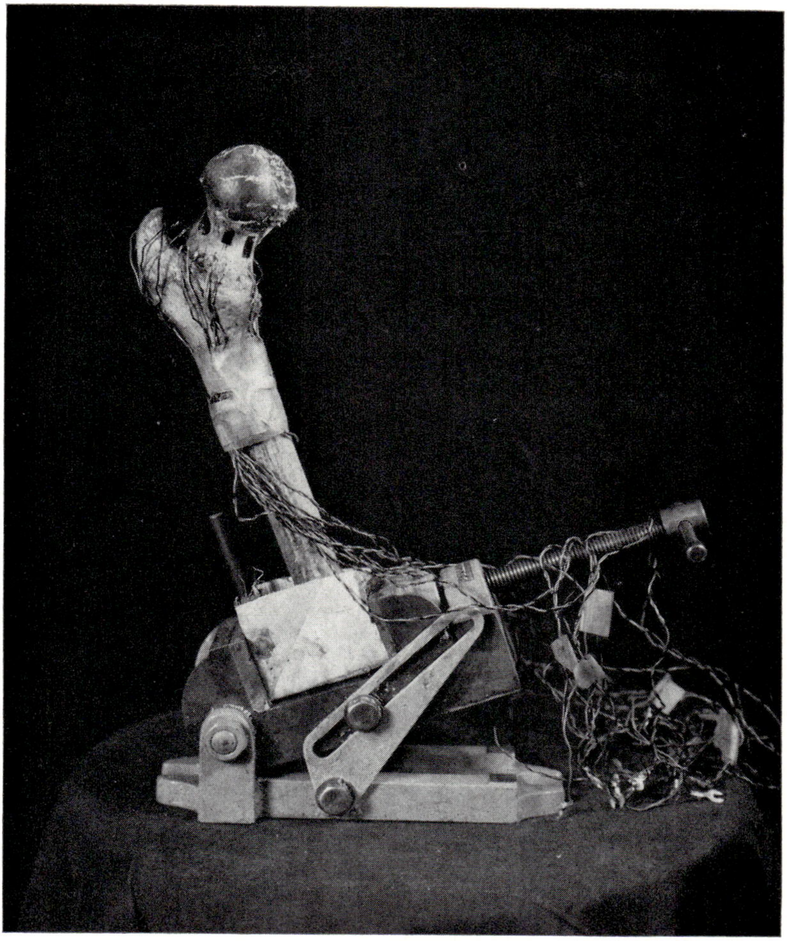

Fig. 2–5. The strain gauge shown is a device used to measure strain. Here it is applied to the neck of a femur in an experimental investigation.

lacquer (Stresscoat*) is sprayed onto a dry surface which is to be strained. When the underlying surface is elongated beyond a critical threshold, cracks appear in the lacquer. The crack patterns are used in the analysis of the strain pattern (Fig. 2–6).

A third method of measuring strain on the surface of a specimen

* The Magnaflux Corporation.

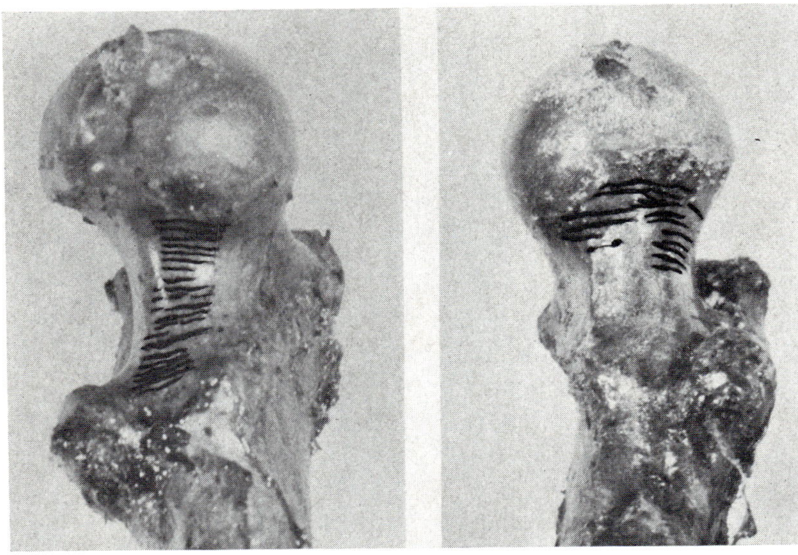

Fig. 2–6. Stresscoat investigation of strain pattern in femoral neck. The strain cracks have been retouched with India ink to make them more visible.

is by the use of a birefringent plastic coating. The coating (Photostress†), whose optical properties vary with the induced strain, is cemented to the surface of the test specimen. When viewed through polarized light, the induced strains are seen as colored fringe patterns. These may be interpreted analytically to determine the underlying strains (Fig. 2–7).

One-dimensional strain may be measured by simple extensometers. These are devices which use either mechanical, optical, or electrical magnification devices to measure small changes in length (Fig. 2–8).

Stresses, with the exception of the surface stress called pressure, cannot be measured since they exist within a material. Photostress and Stresscoat are strain-sensitive coatings and do not directly indicate stress.

If the normal stress on a portion of a body tends to separate this section, the stress is called tension. The simplest case of tension can be illustrated by considering a tendon under the action of a contract-

†Budd Corporation, Instruments Division.

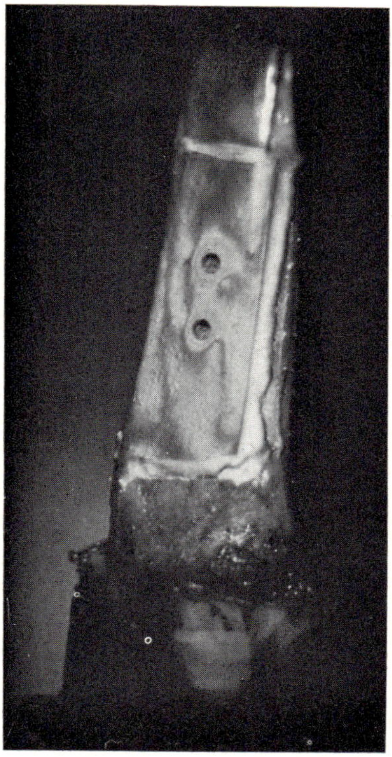

Fig. 2–7. A photograph of fringe patterns around holes in tibia under torsion loading. The fringe patterns may be quantitated with shear strains in the bone.

ing muscle. Using the technique of free body analysis, the stress in the tendon can be determined analytically. If a free body of the section of tendon shown in Figure 2–9 is taken, it shows that the tendon is under a tensile stress, the magnitude of which can be found by dividing the muscle force P by the cross-sectional area of the tendon perpendicular to the force. If the tendon tapers, the stress in the smaller sections will be higher since stress is related to areas in an inverse manner.

The state of strain imposed upon a body subjected to tension is shown in Figure 2–10. It is seen that although the body is lengthened, it also reduces its cross section. This is known as Poisson's effect. Note that in the directions perpendicular to the load, Poisson's effect produces strain. This is true even though there are no stresses on the planes parallel to the load. This should serve to emphasize the

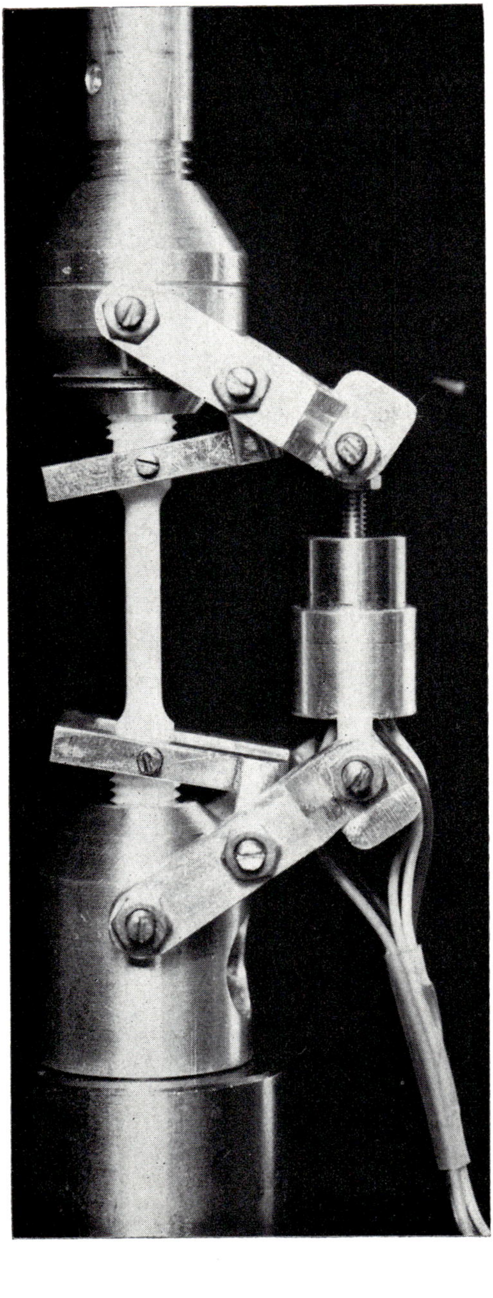

Fig. 2–8. Extensometers are used to measure strain in a bone specimen under tensile loading. The Extensometer consists of a mechanical linkage which carries the strain leading to an electromechanical transducer known as a linear voltage differential transformer.

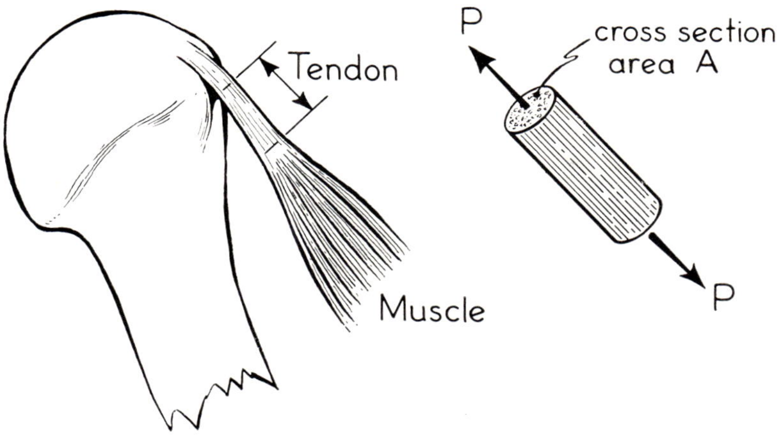

Fig. 2–9. Free body diagram of tendon. The tensile stress is equal to P/A. P is muscle force. A is the cross-sectional area of the plane of interest.

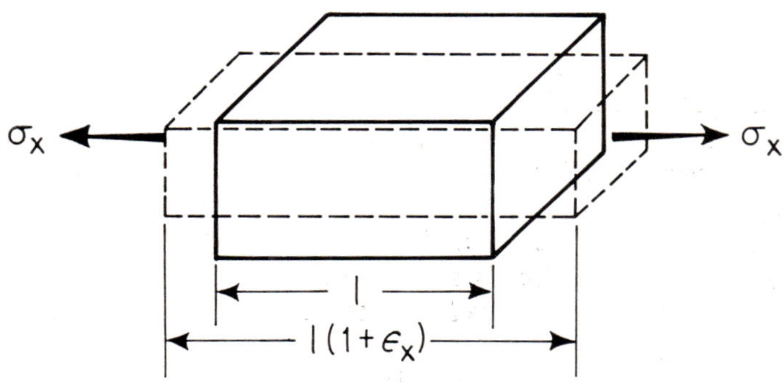

Fig. 2–10. The state of strain induced by simple tensile loading. The body is lengthening, and at the same time, reducing its cross-sectional area. l is the original length. ε_x is the tensile strain. σ_x is the stress produced by the tensile load.

fact that stress and strain are two different physical quantities and that they are not always related on a simple proportional basis.

Compression is the state of normal stress in which portions of the body tend to crush together under the load. Structures such as bone are developed to withstand compressive loading. In a manner similar

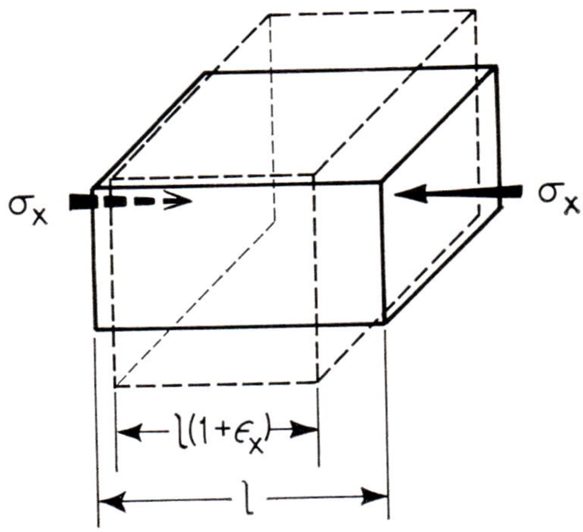

Fig. 2–11. The state of strain introduced in a body by simple compressive loading. The body is decreasing its dimensions in the direction of the load, but increasing the area of planes perpendicular to the level. ε_x is the compressive strain. σ_x is the stress produced by the compressive load.

to the strain condition associated with tension, the strain associated with compression is illustrated by the deformation of a block under load (Fig. 2–11). In this case, Poisson's effect is to expand the dimensions perpendicular to the load. An elongation strain is found in these perpendicular directions without the presence of a tensile stress.

Poisson's effect is a characteristic possessed by all materials. Quantitatively, Poisson's effect is shown in Figure 2–12. Here under the application of a compressive load, the dimension l_1 is changed by an amount Δl_1, which represents the strain in a direction of the load. Simultaneously, the dimension l_2 is increased by an amount Δl_2. This increase in dimension is a strain in the direction perpendicular to the load. The ratio of $\Delta l_2/\Delta l_1$ is Poisson's ratio. In most metals the value of Poisson's ratio is approximately ⅓. The value of Poisson's ratio never exceeds ½. Poisson's ratio is one indication of the compressibility of a material.

In Figure 2–13, the state of strain in both tensile and compressive loadings is shown together with the Poisson's effect. In tensile loading (Fig. 2–13A), the strain in the x or horizontal direction, is ϵ_x; however, the strain in the direction perpendicular to the x direction is $-\nu$ times ϵ_x when ν is Poisson's ratio. Again, with the compressive load-

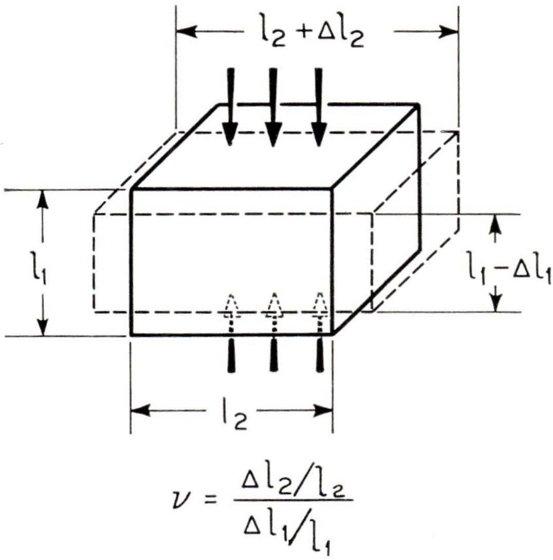

Fig. 2–12. Poisson's effect is defined as the ratio of the induced strain in a direction perpendicular to loading to the induced strain parallel to the loading. Δl_1 is the change in the length of the object in the direction parallel to loading. Δl_2 is the change in length of the object perpendicular to the loading direction. $\frac{\Delta l_2/l_2}{\Delta l_1/l_1}$ as defined is Poisson's ratio.

ing shown in Figure 2–13B, the fore-shortening in the direction of loading is equal to ϵ_x. Note that in this case ϵ_x would be a negative number, indicating a decrease in dimension or compression. The change in the lengths perpendicular to the loading are again $-\nu$ times ϵ_x. Thus, the strains in the directions perpendicular to loading are equal to $-\nu$ times the strains in the direction of the loading.

Stress is a useful concept in studying the strength of a material. It allows the use of the laws of statics and dynamics in the solution of problems concerning the strength of structures such as skeletal members and implants.

Stresses, unlike strains, are not measurable. It is therefore necessary to relate stress to strain empirically. This is accomplished by a series of tests which subject a material of uniform geometry to simple one-dimensional loading. Assumptions are made regarding the uniformity

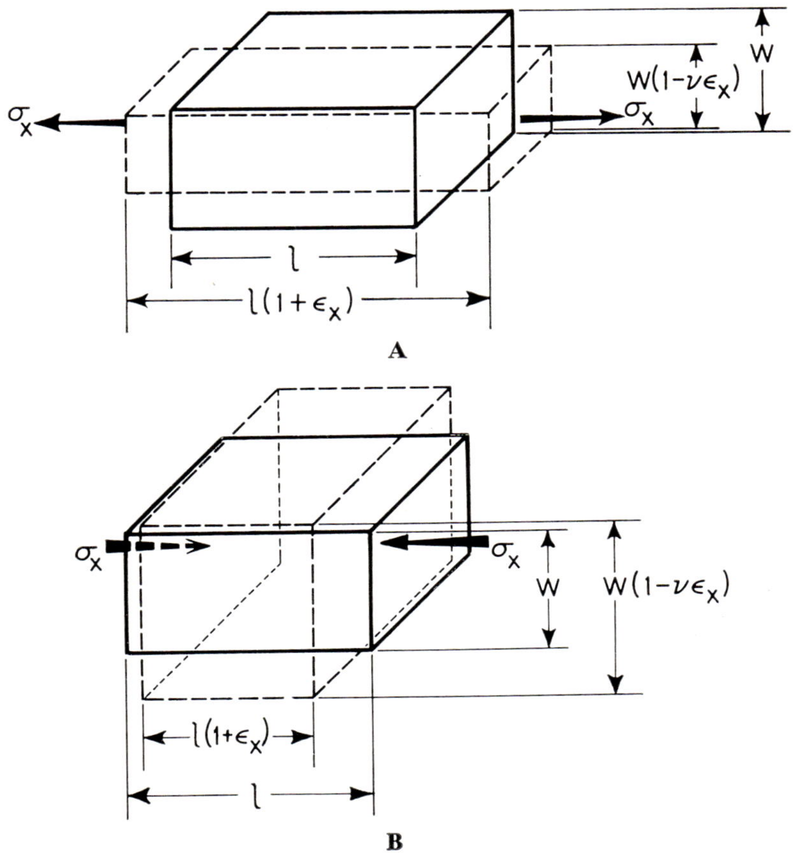

Fig. 2–13. The complete picture of the induced strains under both simple tensile (**A**) and compressive (**B**) loadings. The strains in the directions perpendicular to the loading are equal to Poisson's ratio multiplied by the strains in the directions of the loading. ε_x is the strain in the direction of the loading. σ_x represents the stress induced by the externally applied load. W is the dimension perpendicular to the loading dimension.

of the distribution of the induced stresses and the magnitude of these stresses is then calculated utilizing the concepts of free body analysis and the laws of equilibrium. The strains induced in these loading conditions are measured directly. Empirical relationships can now be formulated relating the calculated stresses and the measured strains.

A typical tension test specimen is shown in Figure 2–8. Tensile loads are applied at the threaded section, and the elongation as well as the change in diameter is measured. Typical data for steel are shown in Figure 2–14.

In the lower portions of the graph the results are almost perfectly linear. The ratio of the change in stress to the change in strain in the direction of the load has been found to be practically constant for many metallic materials independent of the size of the specimen. This ratio of change in stress to change in strain is called the modulus of elasticity. This is represented by the letter E; mathematically, it is written as

$$E = \frac{\sigma}{\varepsilon}$$

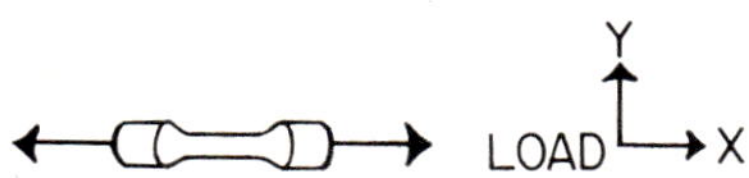

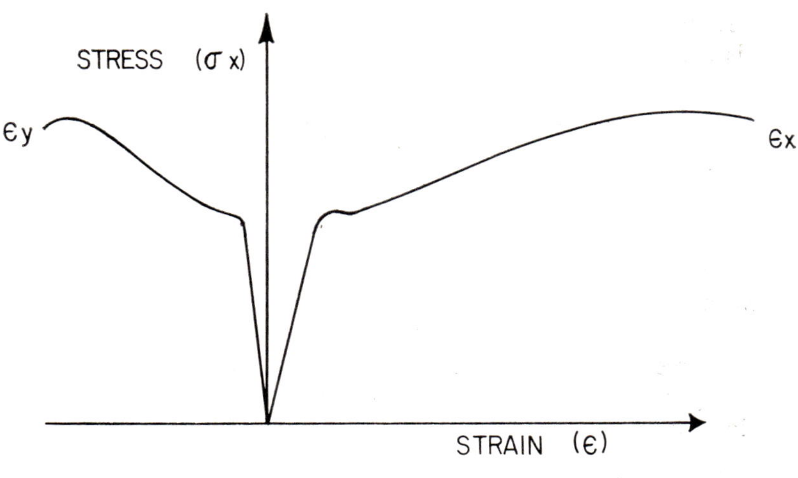

Fig. 2–14. Stress strain curve for mild steel. ε_x is the strain in the direction parallel to the loading, that is, along the axis of the specimen. ε_y is the strain in the direction perpendicular to the axis, that is, in the direction of the diameter of the specimen. σ_x is the tensile stress in the plane perpendicular to the long axis of the specimen.

Since the units of stress are kg/mm², and those of strain are mm per mm, the units of E, the modulus of elasticity, are kg/mm².

Note again in Figure 2–14 that the change in the diameter of the specimen is also linear and is always a fraction of the change in the length of the specimen. The ratio of the change in diameter to the change in length is seen to be Poisson's ratio. It is constant for a given metallic material.

The algebraic expression of these results can be stated as

$$\varepsilon_x = \frac{1}{E}\,\sigma_x$$

and

$$\varepsilon_y = \frac{\nu}{E}\,\sigma_x$$

in which

$$\sigma_x = \frac{P}{A}$$

P is the applied load.
A is the cross-sectional area.
ν is Poisson's ratio.

If the material is isotropic, that is, if its material properties are not dependent on the orientation of the material in the test specimen, the above formulations may be extended to the more realistic case of two-dimensional loading on a three-dimensional body. If a three-dimensional body such as a small cortical bone section is placed under two-dimensional loading, it is possible to determine the induced stresses by measuring the strains. A method of analyzing the relationships between stress and strain is first to consider the strains imposed by the forces in the "x" direction and then to consider the strains imposed by the forces in the "y" direction. The resulting strains will then be superimposed to determine the composite strain configuration. Due to the load in the "x" direction, a stress σ_x is produced. The strains accompanying this stress are:

$$\varepsilon_x = \frac{1}{E}\,\sigma_x$$

$$\varepsilon_y = \frac{-\nu}{E}\,\sigma_x$$

and

$$\varepsilon_z = \frac{-\nu}{E} \sigma_x$$

The σ_x stress is responsible for strains in the three perpendicular directions. In addition to the strains given above, the load in the "y" direction will produce independently the following values of strain:

$$\varepsilon_z = \frac{-\nu}{E} \sigma y$$

$$\varepsilon_x = \frac{-\nu}{E} \sigma y$$

and

$$\varepsilon_y = \frac{1\sigma}{E} y$$

Therefore, the addition of these results to the previous results are:

$$\varepsilon_x = \frac{1}{E} (\sigma_x - \nu\sigma_y)$$

$$\varepsilon_y = \frac{1}{E} (\sigma_y - \nu\sigma_x)$$

$$\varepsilon_z = \frac{-\nu}{E} (\sigma_x + \sigma_y)$$

These algebraic equations may be put in a different form by solving for the stresses in terms of the strains, thus:

$$\sigma_x = \frac{E}{(1 + \nu^2)} (\varepsilon_x + \nu\varepsilon_y)$$

and

$$\sigma_y = \frac{E}{(1 + \nu^2)} (\varepsilon_y + \varepsilon_x)$$

This is the relationship which must be used to calculate the stresses from measured strains. Several things will be apparent. First, in this simple case, in order to determine the stress in a particular direction, strain in two directions must be measured. Second, the value of the modulus of elasticity, E, must be predetermined experimentally. It is important to realize that in studying structures under load, stress and strain do not have a direct one-to-one correspondence.

When two forces which are almost colinear act in opposite directions on a body, their internal effect is to cause a shear. Consider an external force applied to a femoral condyle as shown in Figure 2–15. If the free body of the section of the femoral condyle distal to plane a-a is considered, it is seen that a force F acting parallel to the plane must exist. In addition, there is a moment because the force is acting at a short distance from the plane. This moment is equal to F times d. However, if section a-a is taken very close to the applied force, this moment is negligible and may be disregarded. Thus, force F produces a shear stress equal to force F divided by the area of plane a-a. If the stress is great enough, the section will shear off, producing a typical osteochondral fracture.

Another occurrence of the state of shear stress is found in fluids. Whenever a fluid is flowing, it is subjected to a shear stress. This process can be seen by considering blood as it flows through a vein

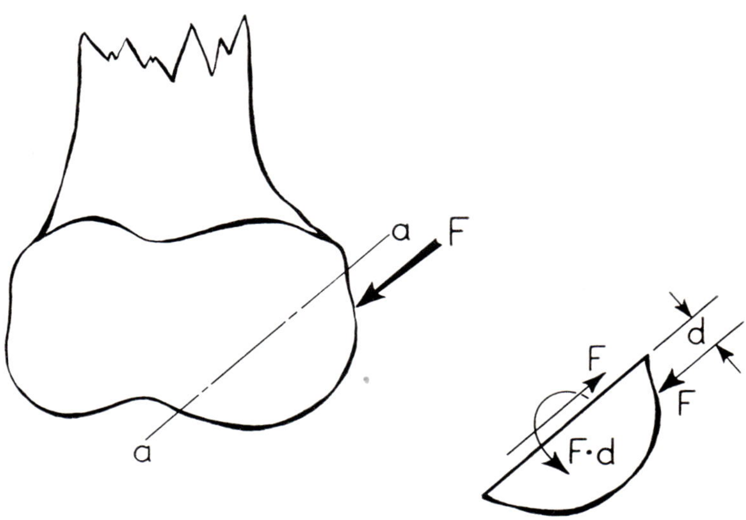

Fig. 2–15. Free body of a section of the femoral condyle under the application of a load F. a-a is an imaginary cutting plane taken parallel to the applied force F. d is the perpendicular distance between the cutting plane and the force F.

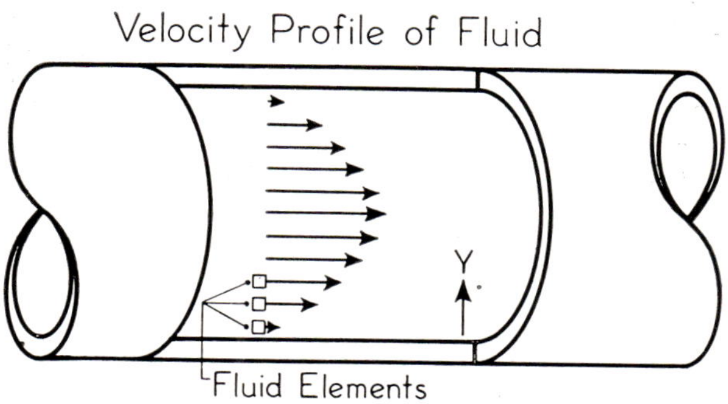

Fig. 2–16. The velocity profile for a fluid moving within a confined tube.

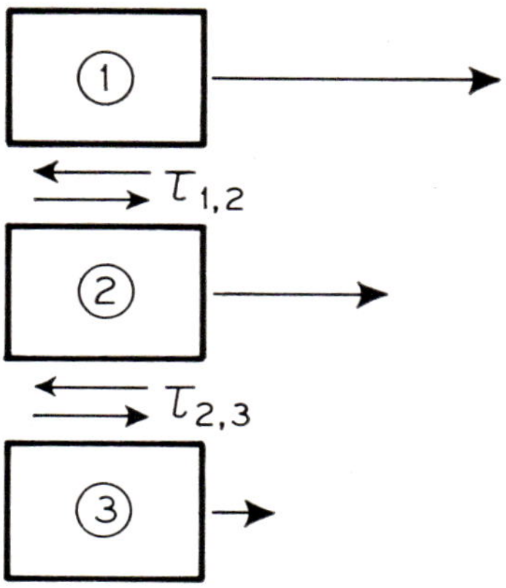

Fig. 2–17. The drag forces on a slug of fluid caused by the differences in relative velocities of adjacent fluid particles. $\tau_{1,2}$ is the shear stress created between fluid slugs 1 and 2; while $\tau_{2,3}$ are the shear stresses created between slugs 2 and 3.

(Fig. 2–16). The fluid at the walls of the vein is not moving due to frictional forces at the wall or molecular attraction at the wall-blood interface. The fluid in the remainder of the vessel has a certain velocity profile. Different portions of the fluid are moving at different velocities. If an element of fluid is considered, it will be realized that the fluid above the element is going faster than the element, while the fluid below the element is traveling at a slower rate. The arrows in Figure 2–17 demonstrate absolute velocities, and it is seen that adjacent elements of fluid have different velocities. The fluid above the element considered is causing a frictional drag on the upper surface, while the fluid below the element is causing the same type of frictional drag on the lower surface, but in the opposite direction. Thus, the drag forces on the element produce shear stresses. An element of blood such as a red blood cell will have shear stresses acting on its surface due to flow in the blood stream. High shear stresses have been suggested as a mechanism of hemolysis.

If a body is subjected to shear stresses and is to be in equilibrium, shear stresses must exist on all faces, as shown in Figure 2–18. This can be most easily understood by consideration of the rotational equilibrium of the body. The sum of the moments must be zero at any point. The stresses in the vertical planes are necessary to counterbalance the stresses in the horizontal plane. When a body is in a state of pure shear, it deforms as illustrated in Figure 2–19. When a shear stress is applied to a block that was originally a cube, it is deformed into a solid parallelogram, and the angles are no longer right angles. This, according to the definition given above, is the shear strain.

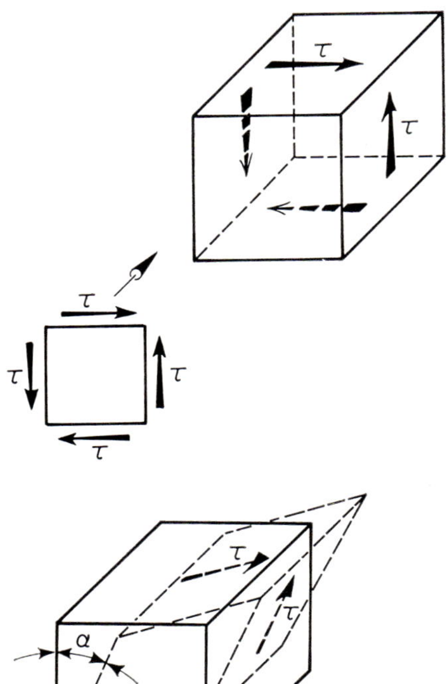

Fig. 2–18. A body in equilibrium subjected to shear stresses must have stresses on all four faces of the body.

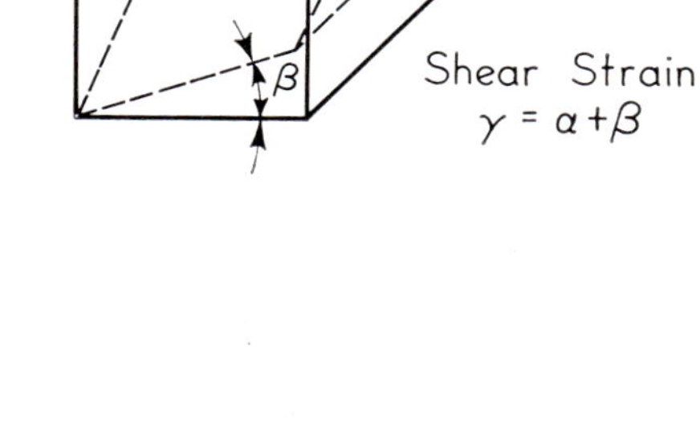

Fig. 2–19. The state of strain imposed upon a body by the application of shear stresses on its surface. α and β are the angular rotations of the planes which were originally vertical and horizontal. γ the shear strain is equal to $\alpha + \beta$; τ is the shear stress.

The modulus of rigidity is the ratio of shear stress to shear strain. It is represented by the letter G. G is found empirically to be equal to τ divided by γ. The units of modulus of rigidity are kilogram per mm^2, or pounds per square inch. Just as the modulus of elasticity represents a measure of the stiffness of the material under tension or compression stress, so the modulus of rigidity is a measure of the stiffness of a material under shear stress.

In order to examine the state of stress in a body, incremental elements must be examined. This is a process of taking a small cubical element and considering it to be a free body. For example, consider the following general two-dimensional case shown in Figure 2–20. It is important to realize that there are stresses on all planes passing through the body. This can be seen by considering a free body of a portion of the cubic elements, as shown in Figure 2–21. A plane is passed through the body at an arbitrary angle, dividing it into two smaller bodies.

In order for these two half bodies to be in equilibrium, there must

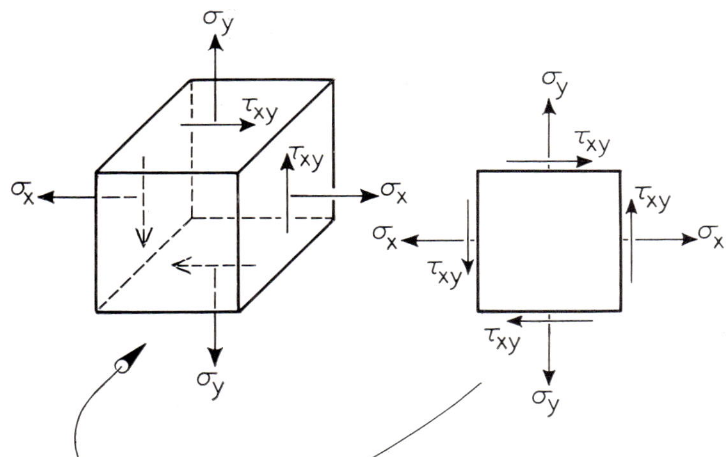

Fig. 2–20. A small cubic element of material, which is subjected to stresses on four of its faces. Each of these faces contains a normal stress, either σ_x or σ_y and a shear stress τ_{xy}. Adjacent to the cube is a representation which may be considered as a flat projection of the cube. Each line in the figure actually represents a surface on the cube.

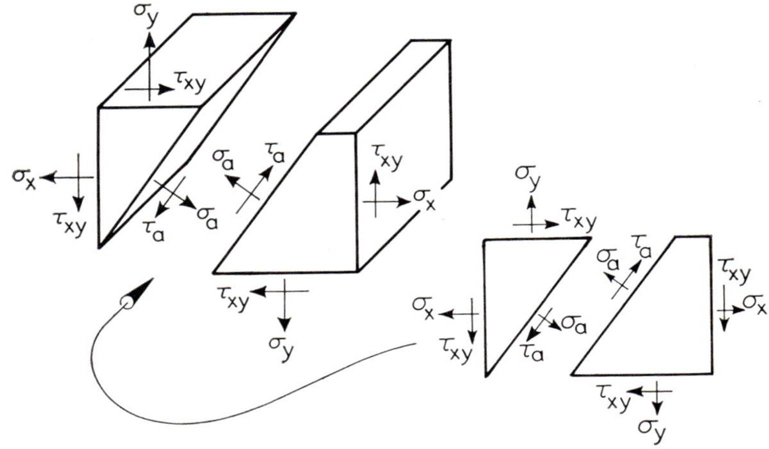

Fig. 2–21. If the small cubic body upon which the stresses are acting is divided into two separate free bodies, stresses of the newly formed surfaces must exist in order to keep the bodies in equilibrium. The notation is carried from Figure 2–20.

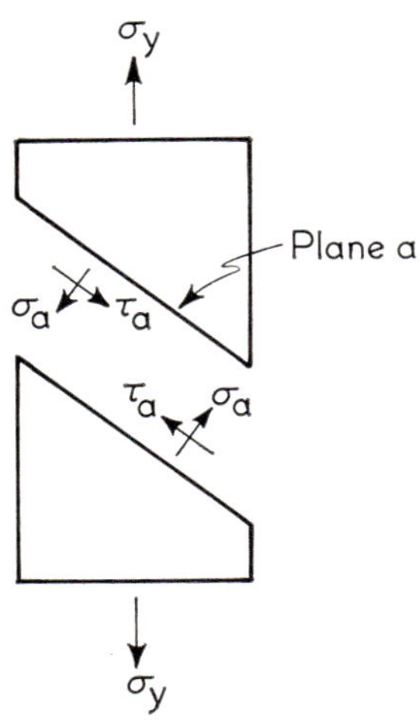

Fig. 2–22. A small cubic body loaded in simple tension. If a cutting plane is passed at an angle of 45 degrees to the direction of loading, then both shearing stresses and normal stresses must act on the cutting plane.

be stresses acting upon the newly cut surface. To visualize this more clearly, let us consider two simpler cases. Figure 2–22 shows a small cubic body loaded only by tensile loading. Shear stresses will be present in an object loaded by simple tension (Fig. 2–22), or compression (Fig. 2–23). Conversely, tension or compression must exist in a body which is loaded by simple shear (Fig. 2–24).

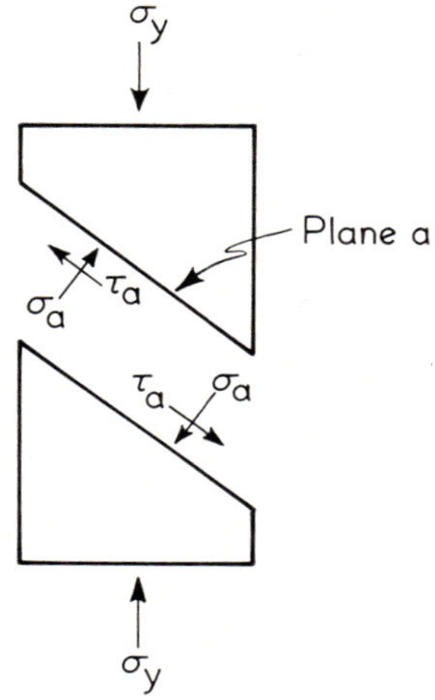

Fig. 2–23. Under the application of simple compressive loading, both compressive and shear stresses must exist on planes oriented obliquely to the line of application to the load.

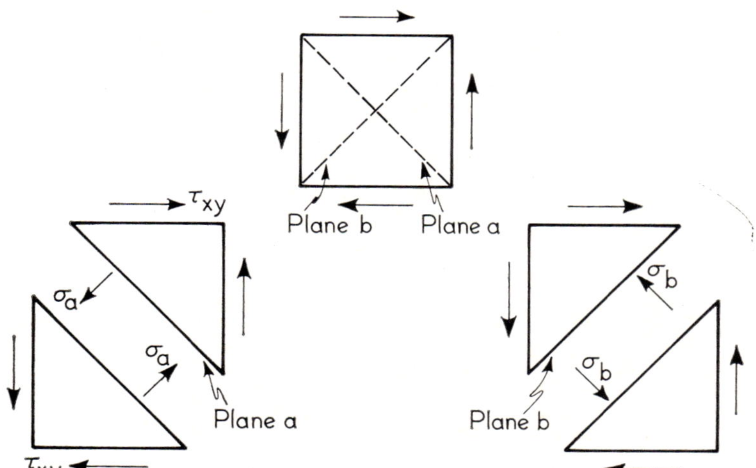

Fig. 2–24. In a body subjected to pure shear, planes passing through the diagonal of the body will be under either pure compression or pure tension stress.

If a specimen is loaded in tension, the plane of maximal tensile stresses, as noted before, will be perpendicular to the axis of the tension load. However, at a plane 45 degrees from this plane of maximal tension stresses, there will be a plane of maximal shear stresses. In Figure 2–25, lines are drawn on the specimen at 45 degree angles to the planes of maximal tension. The figures formed are squares. When the specimen is loaded, the squares deform into parallelograms. It will be recalled that this is a state of shear strain. This is important to realize if one wishes to determine whether a bone or appliance ruptures in a shear or tension mode. Similarly, if the bone specimen is loaded in compression (Fig. 2–26), there will be a plane of maximum shear stress 45 degrees to the plane of maximal compression stress. Note: the square in this figure is deformed into a parallelogram. This indicates shear.

The most important fact to be realized from the above discussion is that a shear stress is present even in an object loaded by simple ten-

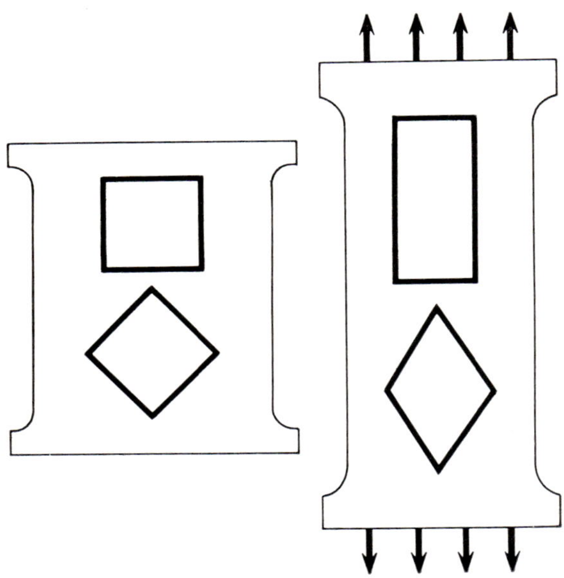

Fig. 2–25. The presence of shear strain in a body loaded in simple tension is graphically illustrated. Lines which were originally at right angles to each other have undergone shear deformation when the load has been applied.

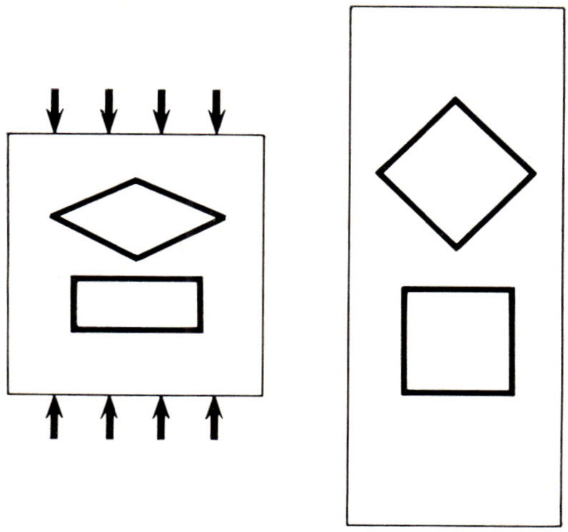

Fig. 2–26. Shear deformations are also present under the application of simple compressive loading.

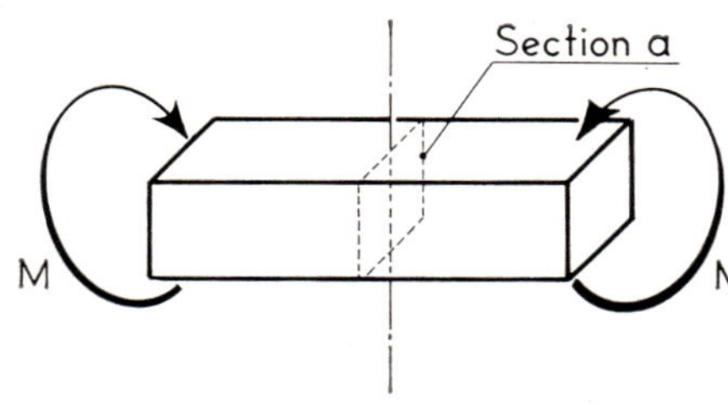

Fig. 2–27. A beam underbending due to the application of moments (M) at each end. Section a is an imaginary cutting plane through the center of the beam.

sion or compression. And conversely, tension or compression must exist in a body which is loaded by simple shear. By noting the inclination of the plane of failure of a structure, it is possible to arrive at some deduction of the mechanism of failure, whether it be shear, tension, or compression.

Up to this point, simple states of stress and their corresponding strain situations have been presented. These must be understood before the more complex but common states of stress can be considered.

The most common complex stress condition existing in bone and in internal fixation devices is bending. The stresses associated with bending loads are a combination of shear, compression, and tension.

To understand the stress condition caused by bending, a short beam of uniform cross section subjected to a moment on each end can be examined (Fig. 2–27). This beam is being subjected to uniform bending. By taking a free body to the left of section "a," the resultant internal reaction on section "a" can be discerned. If a condition of equilibrium exists, it is evident that the results of the internal reaction in section "a" must be equal to the moment applied to the beam. If this were not true, the body would not be in equilibrium and it would tend to spin. This internal reaction moment is achieved by a stress distribution consisting of tension and compression. In Figure 2–28, it can be seen that there is a distribution of compression and tension in the cross section, with maximum compression occurring on the outermost surface of the concave side of the beam and maximum tension occurring at the convex surface of the beam. A similar situation is found in a femoral neck nail plate, when the nail is loaded at the tip. There will be tension at the superior surface of the nail and compression at the inferior surface of the nail.

The normal stress distribution in bending is linear, and varies from zero at the center to a maximum as the outer material fibers are approached. The maximum stress always exists at the outer fiber.

It is possible to calculate the stresses acting at various locations

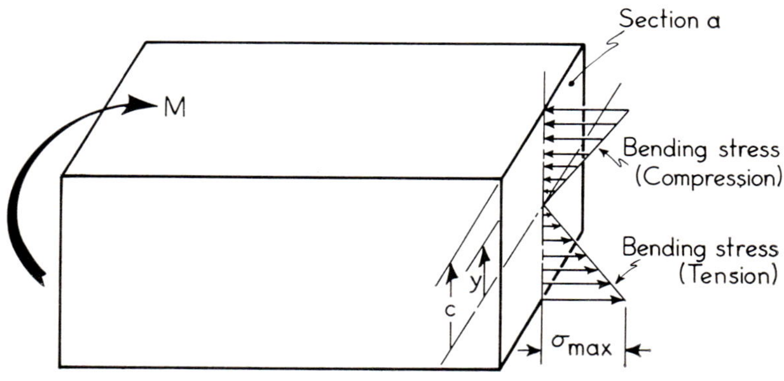

Fig. 2–28. Free body through Section a of beam in Figure 2–27. M is the applied moment. C is the distance from the center of the beam to the outermost fiber. Y is the distance from the bending axis to any particle on Section A of the beam. σ_{max} is the maximum value for tension or compression stress.

throughout the cross section. The stress at any point is related to the bending moment, M, at the section through the point, the distance of the point from the neutral axis, Y, and the area moment of inertia, I_a. The moment of inertia is an expression of the distribution of material in the area. It is dependent upon the shape of the cross section. For instance, material arranged as a three-flanged nail will have a much lower moment of inertia than if it were arranged as an I-beam (Fig. 2–29). The I-beam shape places more material farther from the neutral axis. Bending stress may be calculated from the formula:

Stress is equal to bending moment times the distance from the neutral axis divided by area moment of inertia:

$$\sigma = \frac{MY}{I_a}$$

Since Y is at a maximum at the outer fiber, the stress will be a maximum there. Usually a beam breaks under a bending load because of failure of the fiber, with subsequent rupture of the succeeding outermost fibers. It should be noted that the stress distribution shown applies no matter what cross section is considered. If a cross section is symmetrical with respect to the neutral axis, the maximum compressive stress is equal to the maximum tensile stress. If the cross section is asymmetrical with respect to the neutral axis, as for example in the tibia, the maximum compressive stress under bending load may be greater than the maximum tensile stress, as shown in Figure 2–30.

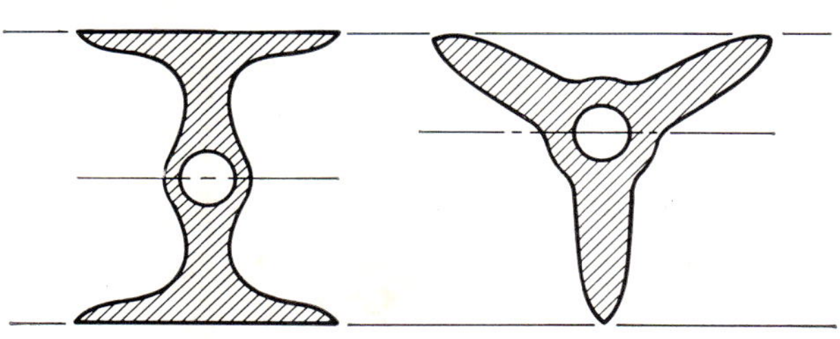

Fig. 2–29. The comparison from the area moments of inertia between an I-beam nail and the flange nail. Both nails have the same cross-sectional area. However, their area moments of inertia are significantly different and the I-beam nail is stiffer and stronger.

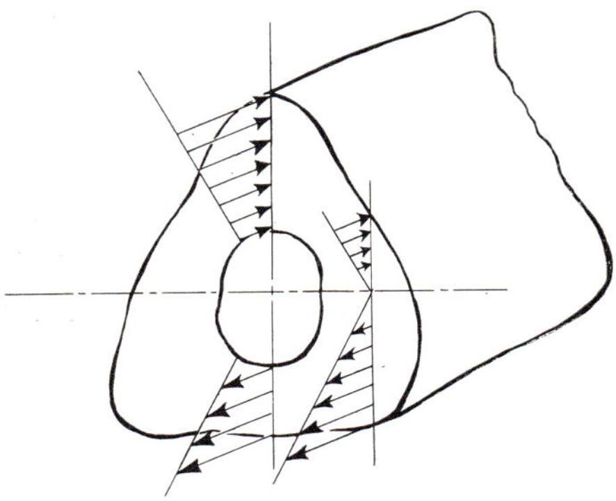

Fig. 2–30. Cross section of a tibia under bending. The shape is not symmetrical with respect to the neutral axis. The bending stresses are still proportional to the distance from the neutral axis and are maximum at outer fibers. Due to the anatomical shape and distribution of material in the tibia, the maximal tensile and compressive stresses may not be equal.

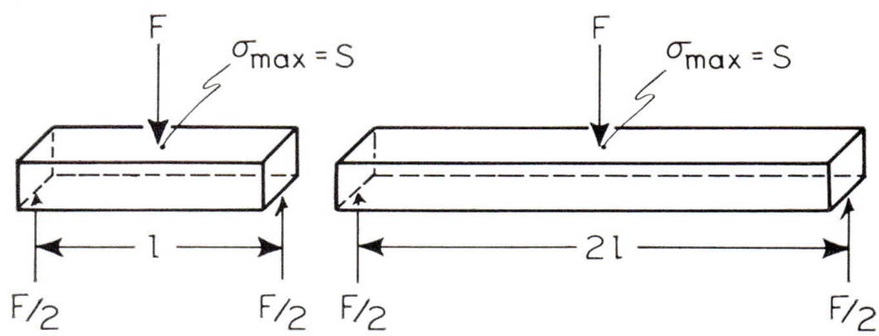

Fig. 2–31. The beam on the right is twice as long as the beam on the left. Thus, the moment at the center of the long beam is twice that of the moment of the center of the short beam. Therefore, the stresses at the outermost fiber at the center of the long beam will be twice that of the maximal stress in the short beam.

Since a knowledge of bending stresses is useful in the understanding of the mechanics of long bones and such internal fixation devices as intramedullary rods, nails, and plates, bending stress will be further examined by studying separately each of the variables: M, the moment; Y, the distance from the neutral axis; and I_a, the area moment of inertia.

The value of the internal bending moment M in any section was shown in Chapter 1 to be the product of a force and a distance. This means that given the same force acting on one end of the beam, the stress at any section distal to the force will vary with the length since the internal bending moment varies with the length. Figure 2–31 demonstrates that for equal sections, a beam that is twice as long will have twice the maximum resulting stress since its maximum moment at its center is twice the value of the shorter beam.

The area moment of inertia is the most important concept in a consideration of the strength of beams. The area moment of inertia is that property of the cross-sectional area that takes into account not only the amount of area, but also the disposition of the area with respect to the neutral axis, or centroidal axis. For a rectangle the moment of inertia is given by the formula:

$$I = \frac{BH^3}{12}$$

where B is the width of the section, H is the height, and the neutral axis is the axis parallel to base B, which passes through the center of gravity of this section. This center of gravity is called the centroid and can be determined for all structures.

Beams may possess the same area but may have widely differing area moments of inertia when the distribution of the material is taken into account. Note in Figure 2–32 that the three beams possess the same area, 4 square inches. Their area moments of inertia

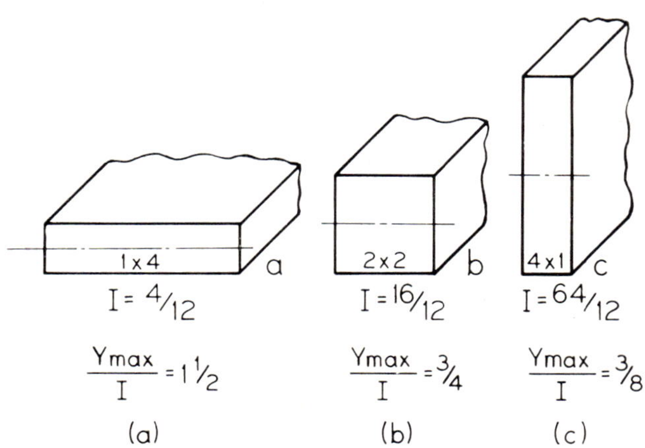

Fig. 2–32. Three beams of equal area. The beam in a has an area moment of inertia of ($4/12$) in.^{4} The beam in b has an area moment of inertia of ($1\frac{1}{3}$) in.^{4} The beam in c has an area moment of inertia of ($5\frac{1}{3}$) in.^{4}

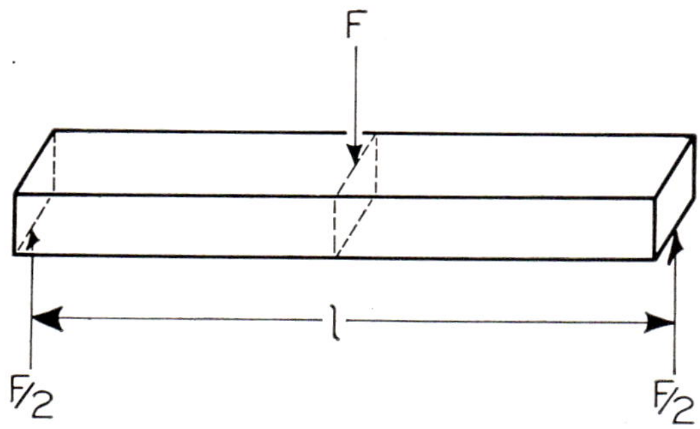

Fig. 2–33. A supported beam loaded in the center by force (F), L indicates the length.

are different owing to the distribution of the material. If the beam is positioned as is shown on the left side of Figure 2–32, it has an area moment of inertia of ⅓ in.[4] If the same area is formed into a square shape, the area moment of inertia will be 1⅓ in.[4] If the beam is turned as is shown on the right of Figure 2–33, the area moment of inertia will be 5⅓ in.[4] There is a factor of 16 between the first and last case. When comparing the strength of these sections of beams in bending, the value of the expression

$$\frac{Y_{max}}{I}$$

is seen to be a measure of the strength of the beam. That is, for a given bending moment the beam which has a minimum value of

$$\frac{Y_{max}}{I}$$

will have the minimum bending stress and will thus have the least tendency to fail.

Therefore, the beam in section a in Figure 2–32 has a value of

$$\frac{Y_{max}}{I}$$

equal to 1.5, while that in section c in Figure 2–32 has a value of $\frac{Y_{max}}{I}$ equal to .375. The beam in section c in Figure 2–32 is 4 times stronger than the beam in section a in Figure 2–32. This indicates that the beam as oriented in section c in Figure 2–32 will support a bending moment which is 4 times the moment which can be supported by the same beam when oriented as in section a in Figure 2–33.

Normal stress may not be the only type of stress produced in a beam under bending. Consider the free body at the left end of the center-loaded beam shown in Figure 2–33. By taking a section of the beam just to the left of the center force, one can see that not only is there a moment at this section, but also, that there must exist a shear force in order for the beam to be in equilibrium as shown in Figure 2–34. This shear force produces a shear stress which is not

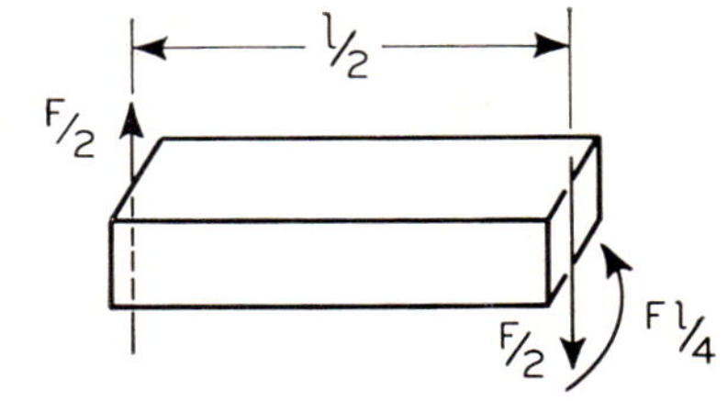

Fig. 2–34. A free body of the left half of the beam in Figure 2–33. l/2 is the length of this portion of the beam. F/2 is the supporting force at the left-hand side. F/2 at the right-hand side is the internal shear force necessary to achieve equilibrium in the vertical direction. Fl/4 is the internal moment necessary to achieve rotational equilibrium.

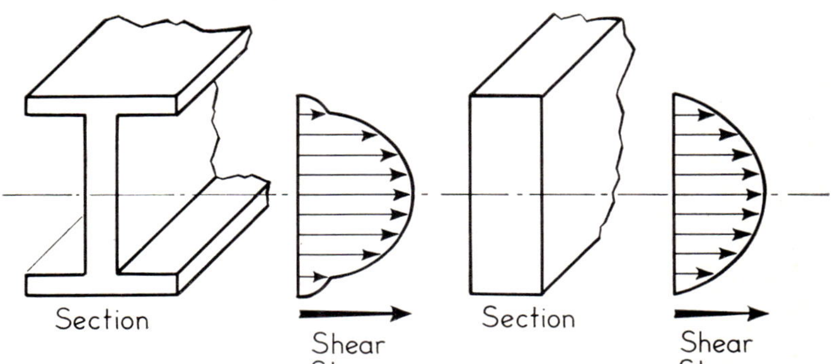

Fig. 2–35. Distribution of shear stresses in two common beam shapes. Shear stresses are zero at the outermost fibers and rise to a maximum at the bending axis.

uniform throughout the section. It is zero on the surface and rises to a maximum at the centrode. Its distribution depends upon the area moment of inertia, and its magnitude usually is much less than the magnitude of the tension or compression stresses arising from the bending moment. In Figure 2–35, two typical shear stress distributions are shown. One is for an I-beam, the other is for a rectangular beam. It should be recalled that the outer surface of an I beam contains those fibers subjected to a maximal normal stress. These are the tensile or compressive stresses, but these fibers also have minimal direct shear stress. For most materials in bending, the bending shear stress is not the critical stress. Its magnitude is usually much lower than that of the bending tension and the bending compression stresses.

The fact that shear stresses exist in any beam that carries loads can be demonstrated by noting the deflection produced in a beam made up of two pieces of wood which have not been attached to each other. Compare this deflection to the kind made of a beam made up of two pieces of wood firmly glued to each other, as shown in Figure 2–36. The glue allows the shear stresses to be resisted by the wood, makes the beam stronger by a factor of 4, and causes it to be more rigid than the beam comprised of the two independent beams. The normal stress distribution in the upper beam varies from tension on the upper face to compression at the lower or contacting face. Similarly, in the lower beam the upper or contacting face is subjected to

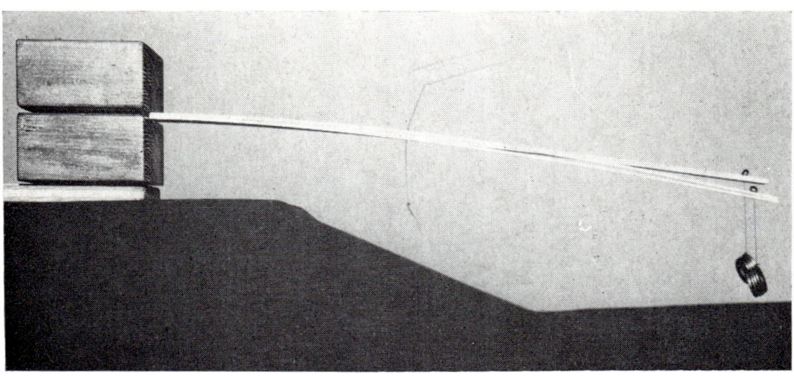

Fig. 2–36. Two beams of equal area. In one, the beam consists of two lamina not attached. The other beam consists of two lamina that have been glued together. I_a for the glued beam is four times that of I_a for the loose beams.

tension, and the lower face is under compression. The value of the ratio of

$$\frac{Y_{max}}{I}$$

for the beam is simply the sum of the independent values. This is in contrast to the behavior of the beam in which the interface has been glued to allow the development of shear stress. The bending stress distribution is tensile on the upper face, zero at the glued interface, and compressive at the bottom face. The value of I is 4 times that of the two single beams and the value of $\frac{Y_{max}}{I}$ is 2 times as high. The glued beam is therefore 4 times as rigid and 2 times as strong as the combination of the two separate beams. Note that without the glued interface there is relative motion between the two surfaces at the center line. If multiple threaded pins are to be used to resist bending loads, they must act in a manner analogous to the laminated structure as illustrated in Figure 2–37. There must be present between the pins osseous material which will allow the formation of bending shear stress. It must be understood that the pin threads will produce sufficient resistance to allow the transmission of the shear stresses from the pins to the cancellous bone. Without either of these two factors, the pin structure will have a strength equal only to the sum of the strength of the individual pins. With these two factors present, the composite structure will have a strength equal to many times the sum of the strengths of the individual pins.

When a body is subjected to a moment or torque in the direction of its axis, it is said to be subjected to torsion. As a result of this applied load the body undergoes a complex state of deformation and stress. Consider the simple case of a solid cylinder subjected to a torque, as shown in Figure 2–38. If the portion of the cylinder to the right of section A is taken as a free body, it is seen that the stress distribution on the face A must produce a resultant moment equal and opposite to the torque imposed on the right boundary of the body. This must be true if this element of the body is to remain in equilibrium. In addition, there must be no resulting axial or radial force, as there is no axially or radially applied load. The shear distribution which satisfies these criteria is shown below in Figure 2–39. It is important to note that all of the stresses in the plane shown are shear stresses.

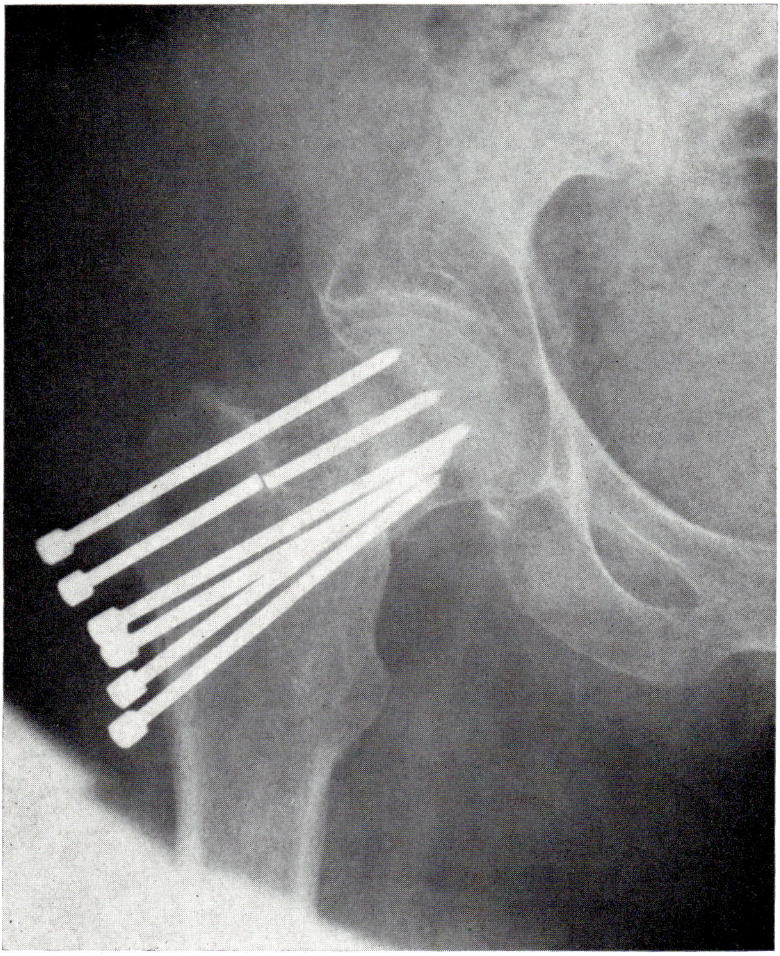

Fig. 2–37. Fracture of the neck of the femur fixed with multiple threaded pins. The multiple threaded pins acted as a single large beam as they were connected by cancellous bone. In a beam of this type, maximal bending shear stress develops in the center as indicated in Figure 2–35. The center pin has undergone a shear failure, indicating that the multiple pins acted as a single large beam. If the pins had acted independently, failure would have occurred by bending of the upper pin.

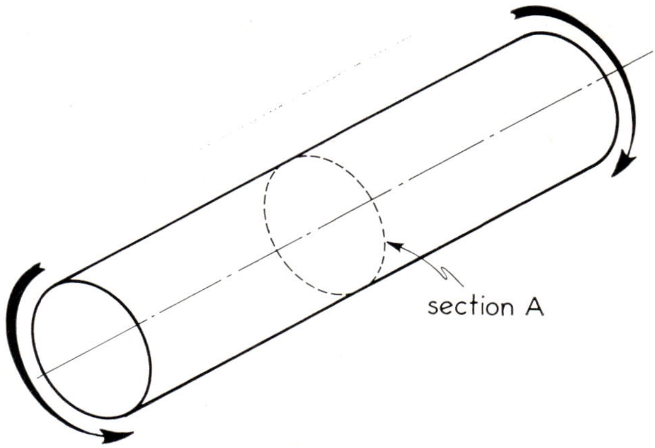

Fig. 2–38. A solid cylinder with an applied torque at each end. Section A is a reference section.

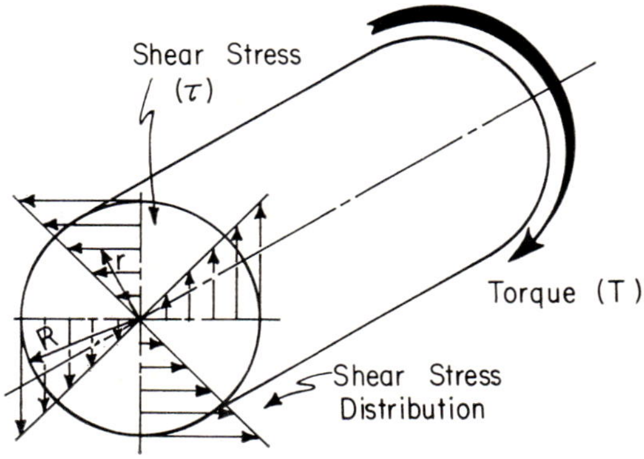

Fig. 2–39. Free body of the right half of Figure 2–38. r is the distance from the center of the cylinder to any point. R is the distance from the center to the extreme fiber. T equals torque. The arrows represent shear stresses (τ).

The shear stress is distributed over the entire area of surface A. The magnitude of the stress is proportional to the distance from the center of the bar. In mathematical form:

$$\tau = \frac{Tr}{J}$$

in which

T is the applied torque,
r is the distance from the center of the cylinder,
J is the polar moment of inertia.

The maximum shear stress will occur at the outer edge (where r = R) and is given by

$$\tau_{max} = \frac{TR}{J}$$

The polar moment of inertia is a quantity similar in nature to the moment of inertia used in bending. It is also a measure of the distribution of the area, but with respect to a central point, not a line as in the case of area moment of inertia. Thus, for a cylinder:

$$J = \frac{\pi R^4}{2}$$

In the case of a human tibia subjected to torsional loading, the stress distribution at two sections is illustrated in Figure 2–40. The shear stress increases linearly with distance from the neutral axis. The value of $\frac{r_{max}}{J}$ varies throughout the length of the bone due to changes in the shape of the bone and the distribution of bony material. At the junction of the proximal ¾ and distal ¼, $\frac{r_{max}}{J}$ is approximately 2 times those at the proximal section even though there is actually more cortical bone at the distal section. Therefore, the stresses at the distal section will be approximately 2 times those at the proximal section. Torsional fractures of the tibia would be expected to occur at the distal section, as is commonly observed. The failure line in a fractured tibia broken in torsion is noted to make an angle of 45 degrees with the long axis of the bone. This can be understood by application of the principles of the state of stress at a point.

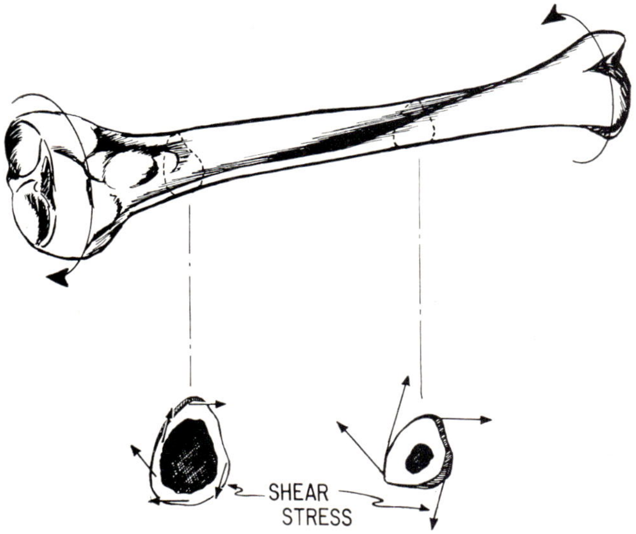

Fig. 2–40. The shear stress distribution at two cross sections in a tibia subjected to a torsional loading. The cross section in the distal end of the tibia, while having more area, has a lower polar moment of inertia; therefore, it suffers greater shear stresses. The section through the proximal third, while having less area of cortical bone, has a more favorable area of distribution, and, therefore, has significantly less shear stress.

Examining a surface element in the region of the fracture of the tibia, we can see that the stress condition is as illustrated in Figure 2–41. Looking at a top view of the element and considering a cutting plane such as B, we see that the stress existing on plane B is tension. The magnitude of this tension stress is equal to the magnitude of the shear stress.

The ultimate or limiting tensile stress for bone under simple loading is less than the ultimate or limiting shear stress; therefore, failure occurs along planes such as B whose tensile stress just exceeds the critical value.

If instead of plane B, plane A is examined, a state of compression will be seen to exist across this plane. The value of this compression is equal to the value of the shear stress. Therefore, in a body under torsional loading, the planes parallel and perpendicular to the axis of loading contain the maximum shear stress. Those planes at angles

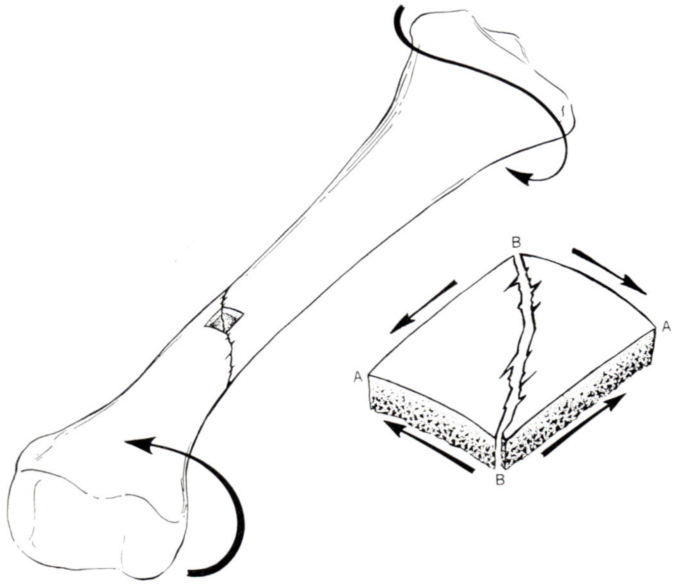

Fig. 2–41. The stress distribution in a small element of the surface of the tibia under torsional loading. If the small element is examined, a diagonal plane such as **B** is seen to have a tensile stress, while a diagonal plane such as A is subjected to a compressive stress.

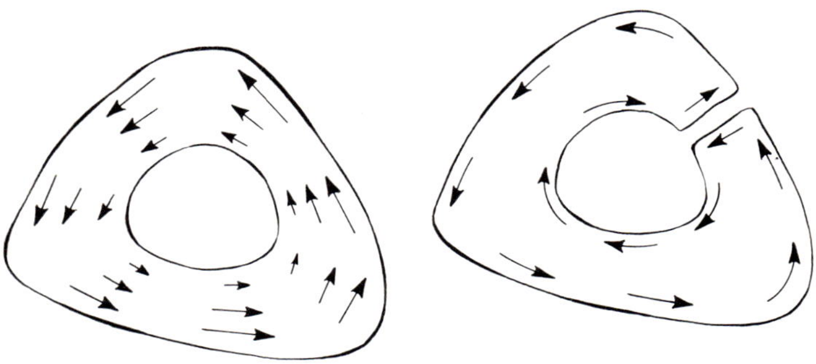

Fig. 2–42. Closed and open sections. The stress distribution produced by torsional loading is shown for each section. In the closed section all shear stresses are running counterclockwise. In the open section only shear stresses on the outer surface are running counterclockwise. In the closed section all shear stresses are effectively resisting the applied torque. In the open section only the shear stresses on the outer surface are resisting the applied torque. As a result the open section can only resist significantly lower torques.

of 45 degrees to the axis contain the maximal tension and compression stresses. Note that according to the previous discussion of the state of stress at a point, there are stresses on all planes in the body but maximal values occur on the planes noted above.

If a ferrous alloy rod is subjected to torsion until failure occurs, the failure plane will be noted to lie perpendicular to the axis of loading, the plane of maximal shear stress. This indicates that the ferrous alloy, unlike bone, fails under simple loading at a lower value of shear stress than of tensile stress.

Objects such as cylinders, hollow cylinders, hollow squares, and tibial cross sections are termed "closed sections." Objects such as I-beams, channel sections, and C sections are termed "open sections." Opening a closed section (*i.e.,* cutting a slot in a tibia) drastically

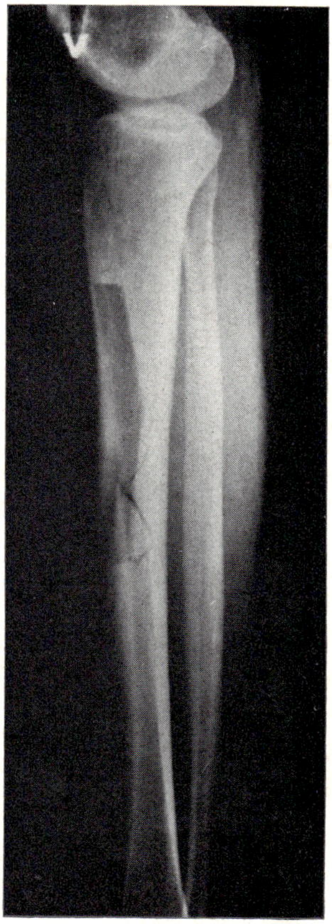

Fig. 2–43. Fracture through a tibial donor site. The patient merely pivoted on the foot.

alters the stress distribution, and hence, the strength of the section. This can be seen by examining the stress distribution in an open section (Fig. 2–42). Note that in the closed section, the stress distribution is such that all stresses have counterclockwise moments about the central axis. Thus, they all effectively contribute to the equalization of the applied torque. In the open section, the torque is equalized because the moment produced by the stresses at the outer surface is slightly greater than the moment produced by the stresses at the inner surfaces. The moment produced by the shear stresses at the inner surface is in the same direction as the applied torque. Therefore, the net amount of torque developed in the section which will resist the applied moment is the difference between the moments produced by the stresses at the outer surface and inner surface. Since the applied torque must be resisted by the induced stresses, very much larger stresses are required in the open section than in the closed section to resist the same torque.

As noted above, the production of shear stresses under torsional loading is associated with the production of tensile and compressive stresses of equal magnitude. In the case of the open section under torsion, large shear, tensile, and compressive stresses are developed in response to the torque. A bone structure fails when the critical tensile stress level is reached. This is noted clinically in fractures of donor tibias or bones changed to open section structures by cysts and tumors (Fig. 2–43).

3. Energy and Work

THE CONCEPTS of energy and work are extremely useful in evaluating the conditions causing fracture in both engineering and biological materials. In order to use these concepts, their definitions must first be discussed.

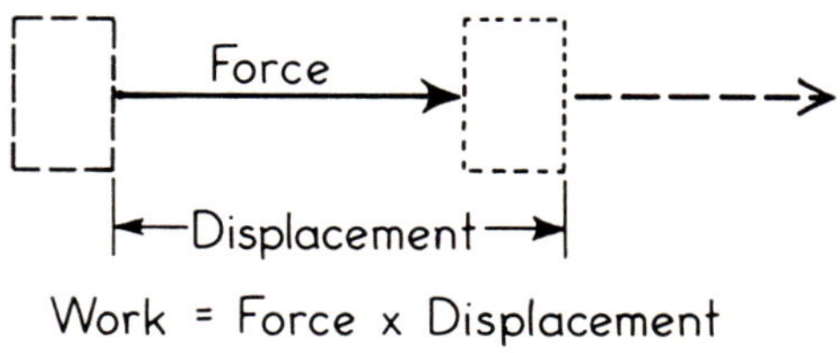

Fig. 3–1A. Work is the effect of a force moving through a distance.

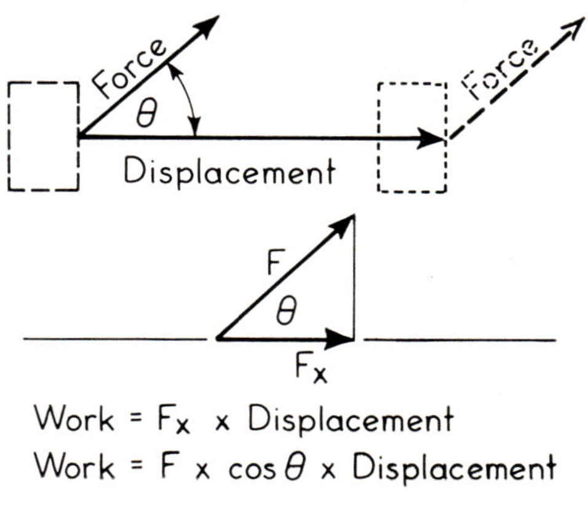

Fig. 3–1B. Force and displacement are vector quantities. The work done is the component of the force in the direction of the displacement multiplied by the displacement. F is the force and F_x is the component of the force in the direction of the displacement. Angle θ is the angle between the force and its component in the direction of the displacement.

Work is the product of a force moving through a displacement. If the force is in the same direction as the displacement, the amount of work produced is (Fig. 3–1A):

$$\text{work} = \text{force} \times \text{distance}$$

Forces and displacements are both vector quantities. Therefore, if the two directions are not the same, the component of the force in the direction of the displacement is utilized in determining the work done, as seen in Figure 3–1B. Since this component is equal to the force multiplied by cosine θ,

$$\text{work} = \text{distance} \times \text{F} \cos \theta$$

Energy is the ability to perform work. There are many forms of energy such as chemical, atomic, thermal, electromagnetic, and mechanical. We shall concern ourselves with three forms of mechanical energy: potential energy, kinetic energy, and strain energy.

Potential energy is the energy possessed by a mass by virtue of its height above a reference datum (Fig. 3–2):

$$\text{PE} = \text{weight} \times \text{height} = 100 \text{ lbs} \times 1.5 \text{ ft} = 150 \text{ ft}\cdot\text{lbs}$$

The mass shown above has 150 foot-pounds of potential energy with respect to the ground level. A 50-kg woman, 160 cm tall, whose center of gravity lies 64 cm above the reference level possesses 2700 kg•cm of potential energy. When she falls, she will lower her center of gravity to the reference level and in doing so will change her potential energy into kinetic energy. The kinetic energy is equal to the potential energy (Fig. 3–3).

Kinetic energy is the energy possessed by a mass by virtue of its velocity relative to a reference mass.

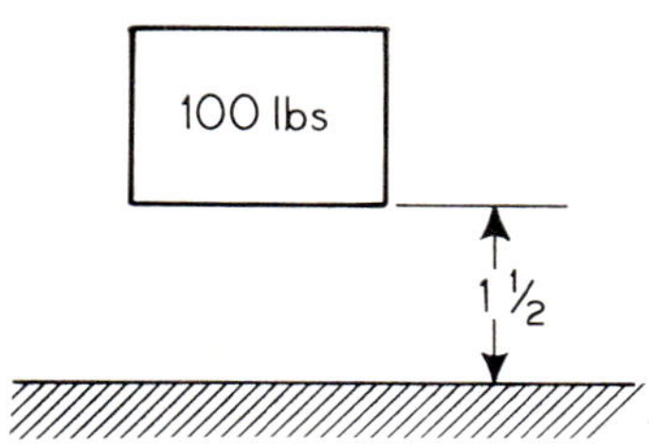

Fig. 3–2. Potential energy. Potential energy is equal to the weight (100 pounds) times the distance (1½ feet), which is equal to 150 foot-pounds.

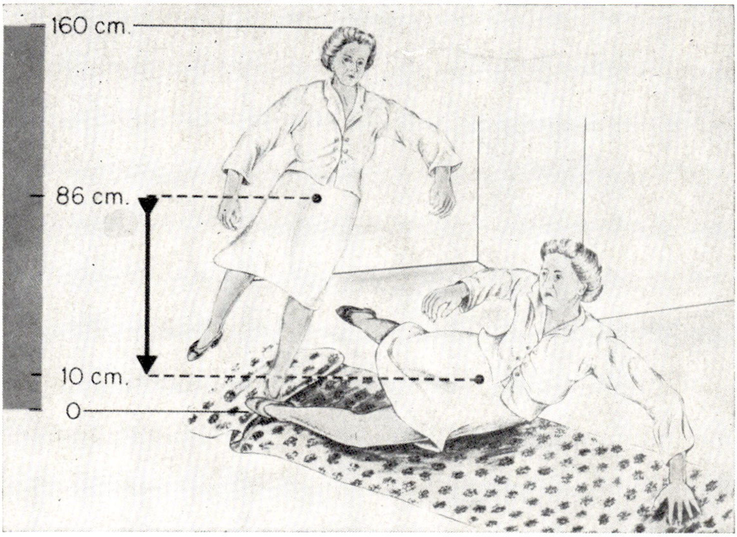

Fig. 3–3. Change in vertical position of the center of gravity of a 50-kg woman during a fall on the hip.

The formula for kinetic energy is:

$$KE = \tfrac{1}{2}\ MV^2$$

where M is the mass of the body, and V is the relative velocity between the body and the ground.

In Figure 3–4, the small mass possesses kinetic energy with reference to the large mass. Thus, in the above illustration, the small body can give the larger body the following kinetic energy.

$$KE = \tfrac{1}{2}\ MV^2$$

$$\text{Mass} = \frac{\text{Weight}}{\text{Gravitational Constant}} = \frac{20\ \text{lbs}}{32.2\ \text{ft/sec}^2}$$

$$\text{Velocity}^2 = (V_{\text{small body}} - V_{\text{large body}})^2 = (100 - 10)^2 = 8100\ (\text{ft/sec})^2$$

$$KE = \tfrac{1}{2}\left(\frac{20}{32.2}\right) \times 8100 = 2520\ \text{ft}\cdot\text{lbs}$$

In the case previously presented of a 50-kg woman who possessed 2700 kg•cm of potential energy, it is possible to calculate the velocity with which she will strike the pavement. This may be done by noting

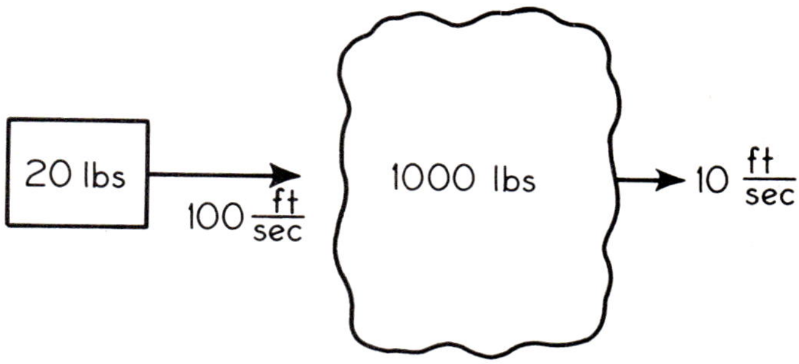

Fig. 3–4. Kinetic energy may be transferred between bodies by virtue of the relative velocity between bodies. The calculations are found in the text.

that the kinetic energy she would possess at the instant of impact must be equal to the potential energy which she possessed at the start of the fall (Fig. 3–3).

Thus

$$KE = \tfrac{1}{2}\ MV^2$$

$$V^2 = \frac{2KE}{M}$$

Since

$$M = \frac{50\ \text{kg}}{980\ \text{cm/sec}^2} = .051\ \frac{\text{kg sec}^2}{\text{cm}}$$

$$V^2 = \frac{2 \times 2700}{.051}$$

$$V = .338\ \text{cm/sec} = 11.1\ \text{ft/sec}$$

$$KE = 2700\ \text{kg}\cdot\text{cm}$$

Strain energy is the energy a body is capable of absorbing by virtue of the body's changing shape under the application of external loads. Consider the process by which energy is put into a beam (Fig. 3–5).

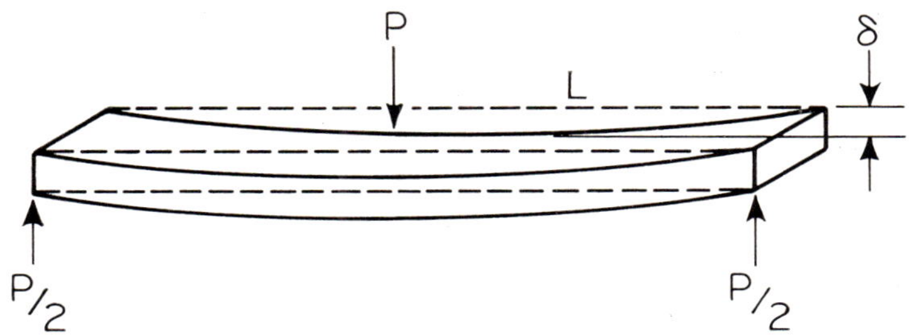

Fig. 3–5. Deflection of a beam; P is the force, L is the length, and δ is the deflection.

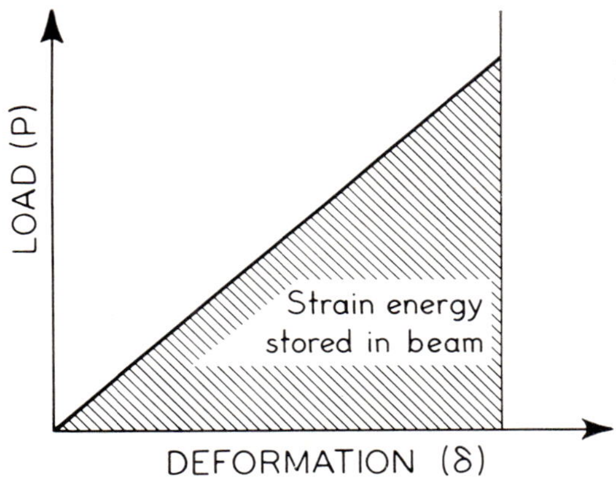

Fig. 3–6. Load-deformation curve for progressive bending of beam simply supported, as in Figure 3-5. The area under the curve represents the strain energy stored in the beam during bending.

As the load P is applied, the deflection δ grows. Actually, the deflection is related to the load by

$$\delta = \frac{PL^3}{48\ EI_a}$$

where

P = force
L = length
E = modulus of elasticity
I_a = area moment of inertia

A load-deformation curve of this process is constructed, as shown in Figure 3–6. The area under the curve represents the strain energy stored during bending of the beam. The work done by the load is given by the formula

$$\text{work} = \tfrac{1}{2}\, P\, \delta$$

which is also the area under the load-deformation curve. All of the work done by the force P on the beam is absorbed by the beam as strain energy.

If the beam is made of a ductile material, and if a sufficient load is placed on the beam, the following load-deformation curve can be obtained, as shown in Figure 3–7.

The left-hand portion of the curve (a-b) represents the familiar reversible linear elastic deflection induced in the material by the bending stresses. At point b the extreme fibers yield at the section with maximum moment. That is, they continue to deform without any increase in stress, and permanent dislocations occur in the crystalline structure. As the loading is continued, the fibers below the

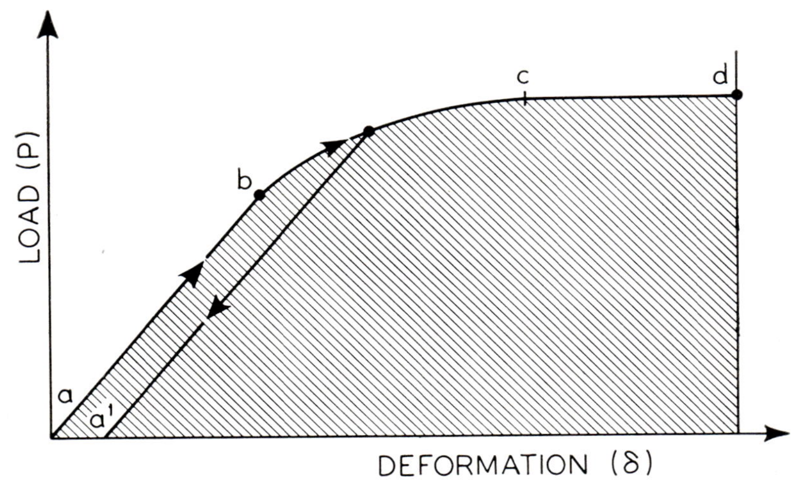

Fig. 3–7. Load-deformation curve for a beam made of a ductile material. The elastic range of the curve is between a and b. At point b, some of the material reaches its elastic limit. At point c, all of the material at a particular section has reached its elastic limit. At point d, the beam ruptured.

extreme fibers yield. If the load is now removed, a permanent change in shape will be noted. If the load is increased, eventually a point is reached (c) at which all fibers at this critical section yield, and, under constant load, the beam continues to deform. This deformation continues until rupture occurs (point d).

The amount of elongation which a simple tension specimen can undergo before rupture is often used as a measure of the ductility of the material. For example, 316 stainless steel elongates 27%, while titanium elongates 36%. Since the elastic strain is only about $\frac{1}{100}$ of the magnitude of the plastic strain, it is apparent that the ability to absorb energy in an implant device is closely related to its plastic deformation. Thus, a nail plate may absorb 0.5 kg•cm before

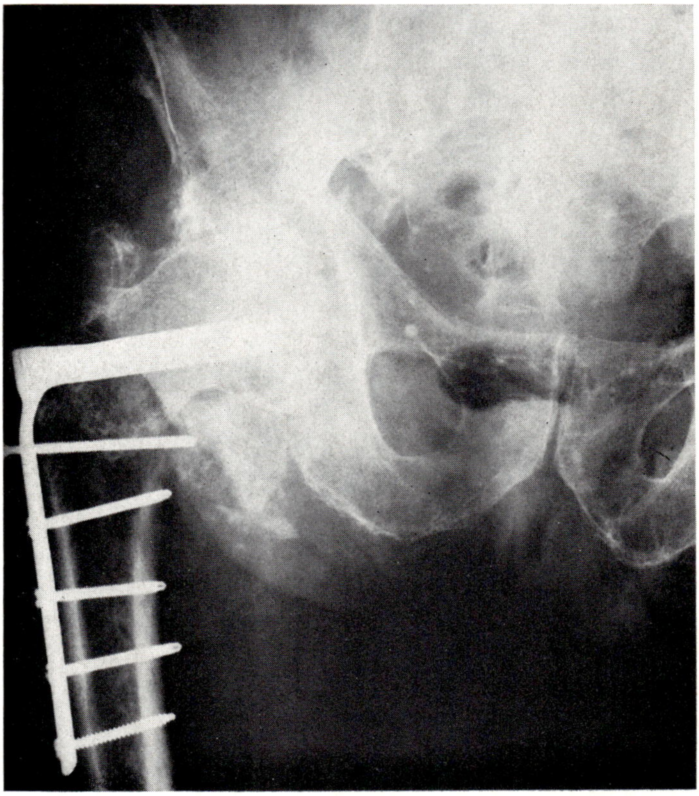

Fig. 3–8. A nail plate which has undergone marked plastic deformation.

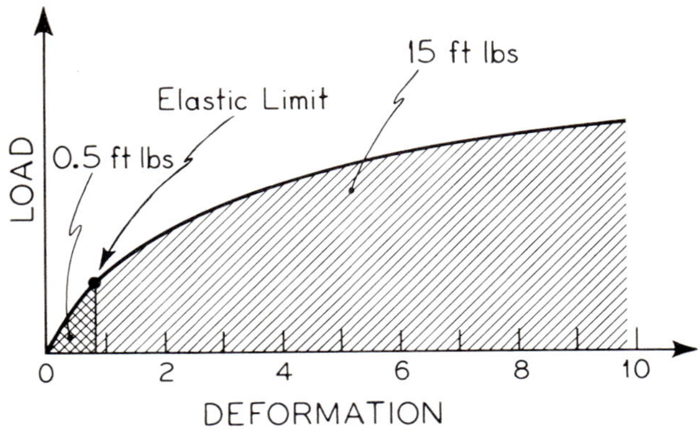

Fig. 3–9. Load-deformation curve for nail plate seen in Figure 3–8. The heavily shaded area represents the energy stored in the elastic range. The lightly shaded area represents the amount of energy that the nail plate can store by deforming.

it incurs a permanent deformation, and before fracture occurs it will absorb 15 kg•cm as illustrated in Figures 3–8 and 3–9.

The energy absorbed to failure of a femoral neck has been found experimentally to be 60 kg•cm. As noted above, on falling a 50-kg woman develops 2700 kg•cm of kinetic energy, which must be dissipated when she hits the floor. This example illustrates the utility of energy concepts in the prediction of failure in mechanical and biological systems in the study of trauma. When two bodies collide, it is difficult, if not impossible, to calculate the forces generated. The energy transfers are relatively simple to determine. In considering the falling woman, 2700 kg•cm of energy must be dissipated. As noted above, only 60 kg•cm of energy can be dissipated by the femoral neck before failure. The remainder is dissipated in the upper and lower limbs by muscle action, elastic and plastic strain energy in the soft tissues and bones, by rolling, and by the strain energy in the floor surface. In Figure 3–10, some of the various methods of energy dissipation are illustrated. Sufficient energy is always present to fracture the neck of the femur when a person falls on the hip. It is the availability of other methods of energy dissipation that prevents fracture of the femoral neck from occurring.

The body possesses many mechanisms for keeping energy absorption within tolerable limits. A child soon learns that a ball will not sting

A, Energy absorption—muscles of lower limb.

B, Energy absorption—muscles of upper limb.

Fig. 3–10. Some methods of energy dissipation during a fall on the hip.

C, Energy absorption—elastic strain on soft tissues of lower limb.

D, Energy absorption—elastic strain on skeletal members of upper limb.

Fig. 3–10. Some methods of energy dissipation during a fall on the hip *(continued).*

E, Energy absorption—elastic strain on skeletal members of lower limb.

F, Energy absorption—plastic strain on skeletal members of upper limb.

Fig. 3–10. Some methods of energy dissipation during a fall on the hip *(continued).*

G, Energy absorption—plastic strain on skeletal members of lower limb.

H, Energy absorption—floor material.

Fig. 3–10. Some methods of energy dissipation during a fall on the hip *(continued).*

his hands if he pulls them back as he catches it and thus spreads the energy absorption over a longer period of time. The karate enthusiast spends hours hypertrophying the soft tissues on the ulnar border of the hand in order to develop a good pad of energy-absorbing soft tissue. Large calluses develop on the feet of those who habitually walk barefooted. The tumbler trains himself to roll and to lower himself to the ground with quick muscle action. Attempts are made to minimize the effects of trauma by wearing energy-absorbing athletic equipment.

When considering failure of skeletal members that have been altered by previous fracture, disease, or surgery, energy considerations are most important. Torsional fractures of the tibia commonly occur as the result of trauma suffered during such athletic activities as skiing. Work is done by the external torque as it rotates the body through an angle. This work, a torque moving through an angle, is momentarily stored as strain energy in the tibia. In Figure 3–11, a torque-rotation curve for a fresh tibia shows that the energy stored before failure equals 131 kg•cm and is the area under the curve. Note

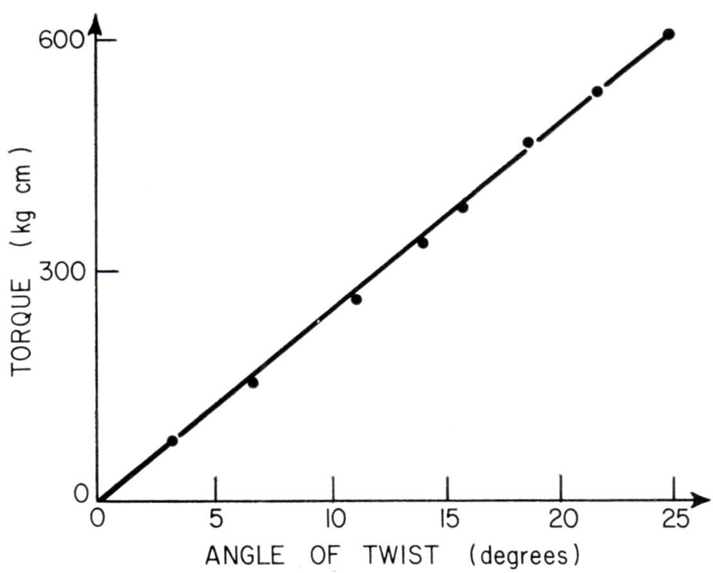

Fig. 3–11. A torque versus angular deflection curve for a fresh tibia. Energy stored is represented by the area under the curve.

that this energy is stored because the tibia rotates through an angle of 25 degrees as the torque of 600 kg•cm is applied.

If the bone is altered by removing a bone graft and converting it to an open section, the torsional load capacity and the obtainable angular deformity are reduced to 1/3 of their previous values (Fig. 3–12). The strain energy stored is reduced to 1/9 of the strain energy that can be stored by the intact bone. While the altered bone can withstand 1/3 of the load of the intact bone, its strength, when viewed as the ability to store energy under conditions of external trauma, is reduced to about 10% of its original value. This explains why bones altered by disease or surgical intervention suffer torsional fractures during essentially normal functions.

The study of strain energy offers an explanation of the weakening effect of a hole. In Figure 3–13 a fracture of the femur is represented. This occurred at the hole left in the bone after removal of a

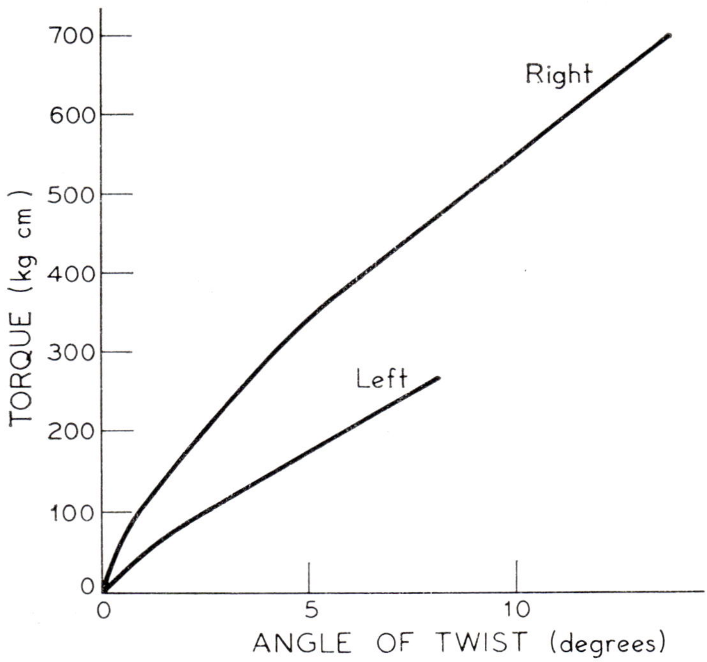

Fig. 3–12. A comparison of the capacity of an intact tibia to absorb strain energy with that of a tibia converted into an open section by removal of a bone graft. *Right,* The intact tibia; *left,* the open section tibia.

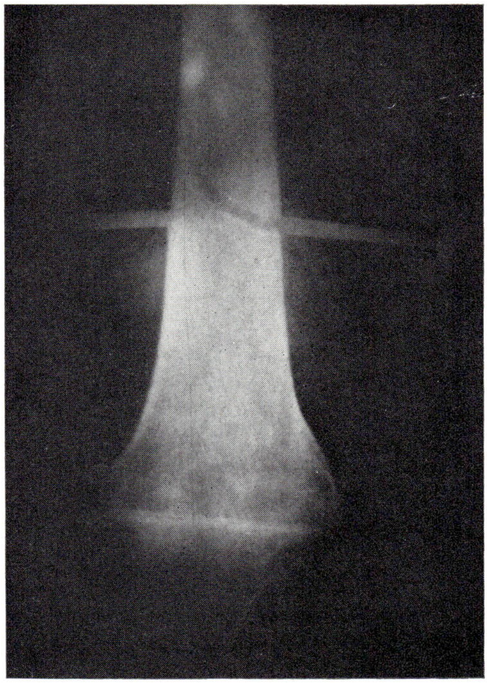

Fig. 3–13. Fracture of the shaft of the femur in a patient with an arthrodesed knee. The fracture occurred after the Steinmann pin used in the compression apparatus had been removed. The fracture line originated in the hole left by the Steinmann pin. The figure is a superposition of x rays taken before the pin was removed and after fracture.

compression apparatus used in an arthrodesis of the knee joint. The trauma was minimal. In the region of a small hole in the cortex of a long bone, there exists a stress-concentration effect when the bone is loaded in any manner. This means that the stresses in the material immediately adjacent to the hole are greater than the stresses in nearby areas. This difference is approximately a factor of 1.6 for a small hole in bone irrespective of its size. If the hole reaches the size of ½ the diameter of the bone, however, the effect is to produce an open section, a phenomenon which was discussed above. In situations in which small holes exist in the cortex, as the load is applied to the bone the stresses in the area immediately adjacent to the hole reach their critical or ultimate value, while the stresses in nearby areas are only ½ of their critical failure value. As can be seen in Figure 3–14, this state occurs at point A halfway up the load-deformation curve. Since the hole does not alter the stiffness of the bone, the deformation corresponding to this load is also one half of the

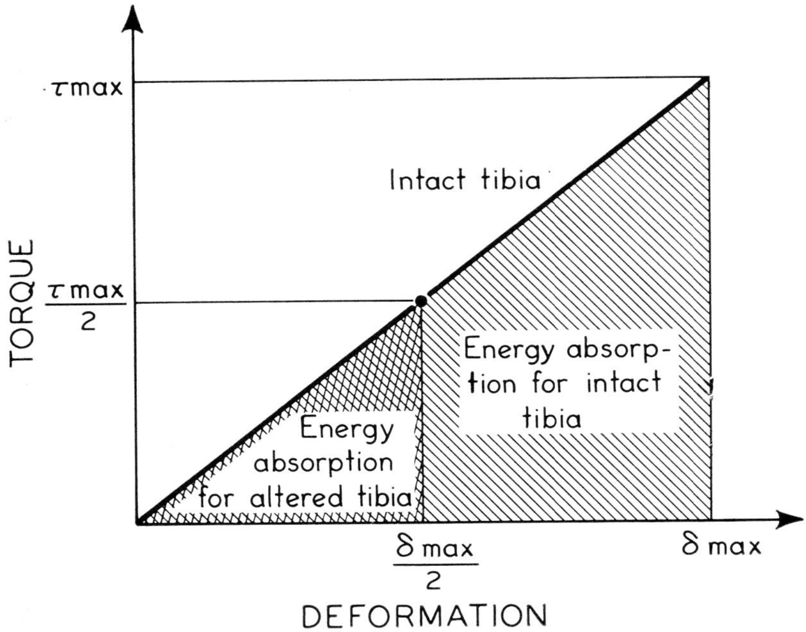

Fig. 3–14. Load-deformation curve for a tibia loaded in torsion. The effect of the hole in this tibia was to cause the stresses to reach the critical failure value at one half of the torque which the tibia could normally carry. The deformation will also be half of the total deformation that an intact tibia could sustain. The total energy that a tibia with a hole can absorb is theoretically ¼ of the energy that a tibia which is intact can absorb.

maximal obtainable value. The strain energy storable in the bone and represented by the shaded area under the curve is one quarter of that storable by the intact bone. If the hole happens to be at a normally critical section of a bone in the region of the minimal polar moment of inertia, for example, failure of the bone may occur with what would be considered negligible trauma.

These energy concepts are useful in understanding failures in columnar structures. A column is a long slender structure loaded axially or a collection of shorter structures elastically supported at their junctions and loaded axially (Fig. 3–15). When a similar structure is loaded as a beam, the amount of energy that can be absorbed varies linearly with the induced deflection (Fig. 3–6). Columns behave

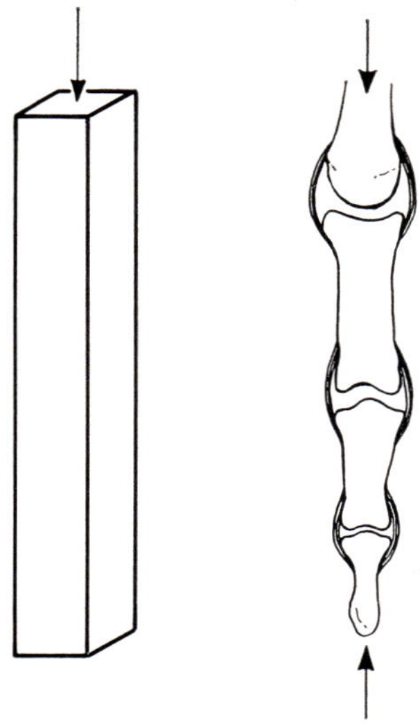

Fig. 3–15. Two types of columns: on the left, a slender elastic member; on the right, a column made up of short elastic segments connected and supported with elastic material.

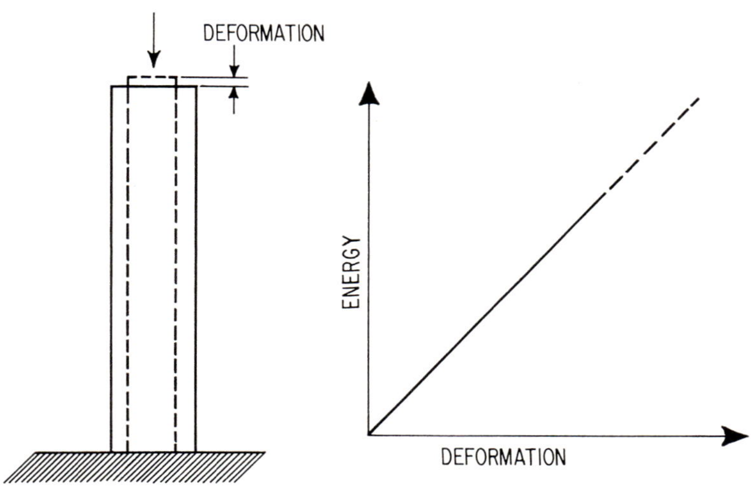

Fig. 3–16. The deformation produced by a small load on the column. The energy-deformation curve shown on the right-hand side of the picture is linear for small deformations.

in a highly nonlinear way. Figure 3–16 illustrates the deformation produced by a small load on the column. The amount of energy absorbed in this manner is the product of the load and the deformation. If the column could remain vertically oriented, the energy would increase with the load in the linear manner of Figure 3–16. There are other configurations, however, that the column may assume. One such is illustrated in Figure 3–17A. The energy-deformation characteristics for this configuration are shown in Figure 3–17B. The initial straight line portion—a-b—represents the erect column being loaded axially and remaining in a linear configuration. In region c there is a marked bow to the column. In region d, the bow is more pronounced, but the total energy is less than in column c. Energy is lost in vibration when the column snaps from one position to another. Actually, if the column reaches the energy storage level indicated at point b, any small disturbance will drastically change its configuration to that at d. The same principles of energy absorption apply to segmented columns. Figure 3–18 shows a drawing from the x-ray

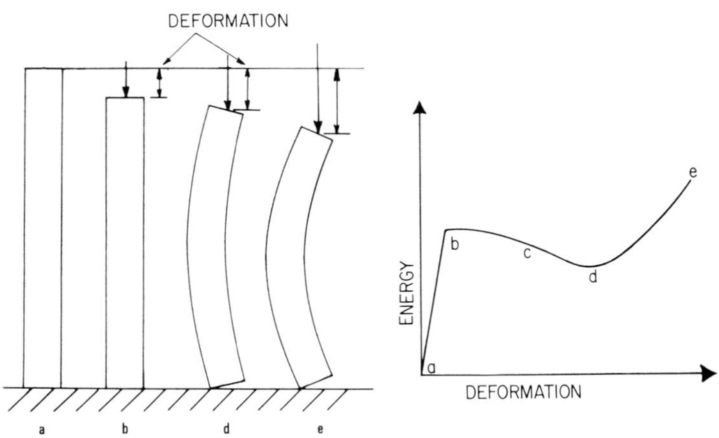

A B

Fig. 3–17. **A,** Configuration assumed by a column under high loads. **B,** Region a-b displays the linear absorption characteristics for small values of a deformation. Once point b is reached, very small amounts of additional energy input are required to produce very large deformations. In region d, a marked deformation appears. To further deform the member such as e, significant additional amounts of energy are required.

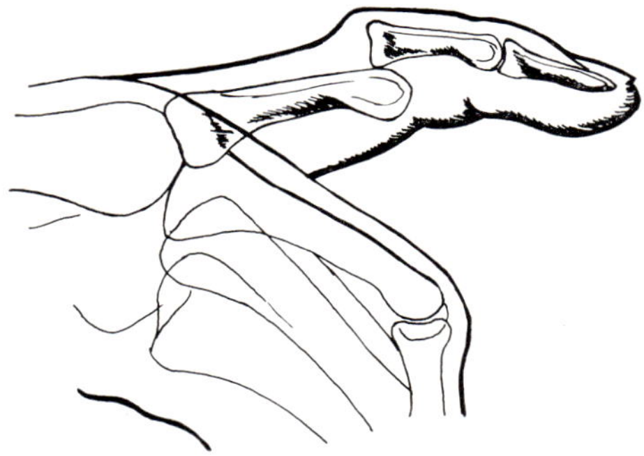

Fig. 3–18. A tracing from an x-ray of a dislocation of the distal IP joint produced as the column buckled in order to absorb the energy of a basketball.

picture of the finger of a 12-year-old girl who was playing basketball and was struck on the tip of the outstretched finger by the ball. She suffered a dislocation at the distal IP joint. The finger acts as a segmented column with elastic constraints at the joint. The column was loaded elastically by the basketball. In order to absorb the excessive amount of energy put into the system, the short segmented column buckled and resulted in a dislocation.

4. Viscoelasticity

BIOLOGICAL materials under load may exhibit a more complex behavior than the materials discussed in the previous chapters. For example, the wrinkles of a bedsheet may replicate their image in the skin after a night's rest. The wrinkles do not disappear immediately, but slowly fade after awakening and beginning the day's activities. If one presses a wrinkle into his skin momentarily, the mark that is produced will fade almost immediately.

It is evident that the duration of the contact pressure is of great importance. It has been demonstrated that earlobe tissue under tension loading exhibits a time-dependent stretch; that is, the material will continue to stretch as a constant load is left on for measurable periods of time. The earlobe can exhibit four well-defined regions in its reaction to the load (Fig. 4–1). As the load is applied, there

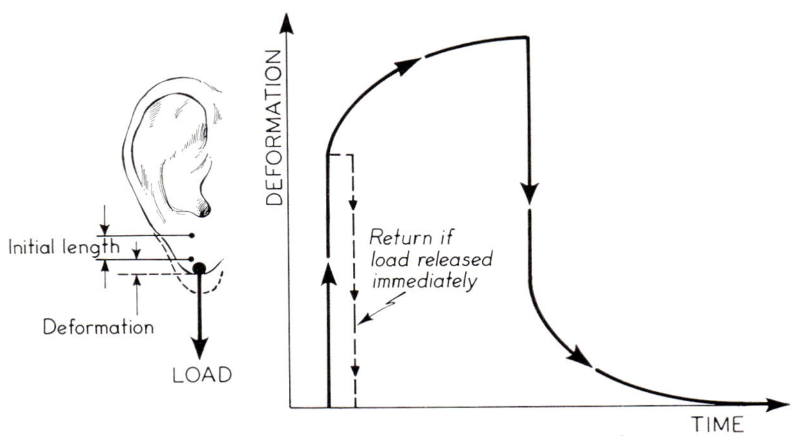

Fig. 4–1. Load-time deformation characteristics of earlobe tissue.

is an area of immediate elastic deformation. In this straight-line region of the curve, if the load is taken off immediately, the material recoils elastically back to its original size. If, however, the load is left on, the material continues to stretch. This stretching tends to reach a limiting or steady state value. If the load is allowed to remain for an appreciable period of time and is then removed, there is an immediate elastic recoil toward the material's original dimension. The material does not, however, reach its original dimension until after a definite period has elapsed. This response to prolonged loading is typical of the response shown by materials classified as viscoelastic.

The above examples demonstrate viscoelastic behavior *in vivo.* Viscoelastic behavior of biological materials has been studied *in vitro.* References to these studies will be made after a consideration of the fundamental concepts of viscoelastic behavior.

The concepts of viscoelastic behavior can be best understood by the construction of an idealized viscoelastic model. One of the two fundamental elements of a viscoelastic model is the Hookean body, as illustrated in Figure 4–2. This body may be characterized as an idealized spring. Its elongation is linearly proportional to the load applied to it. The ratio of the applied load to the change in spring length is called the spring constant. When the load is removed, the body returns to its original shape. In addition, the change in length resulting from the application and removal of a load is instantaneous. The Hookean body does not show any time dependency between applied loads and the resulting deformations (Fig. 4–3).

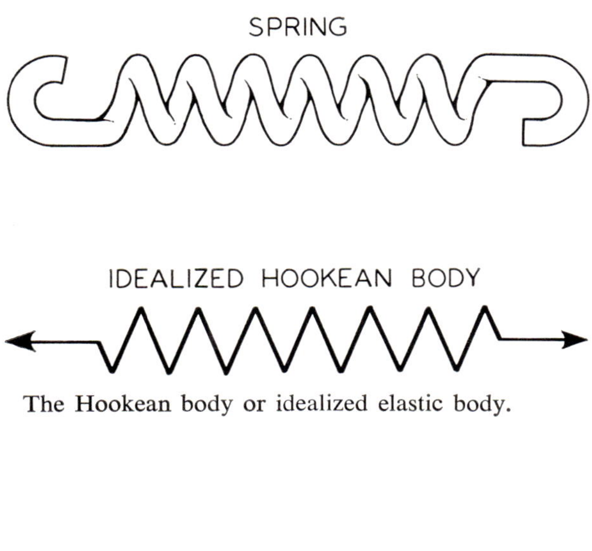

Fig. 4–2. The Hookean body or idealized elastic body.

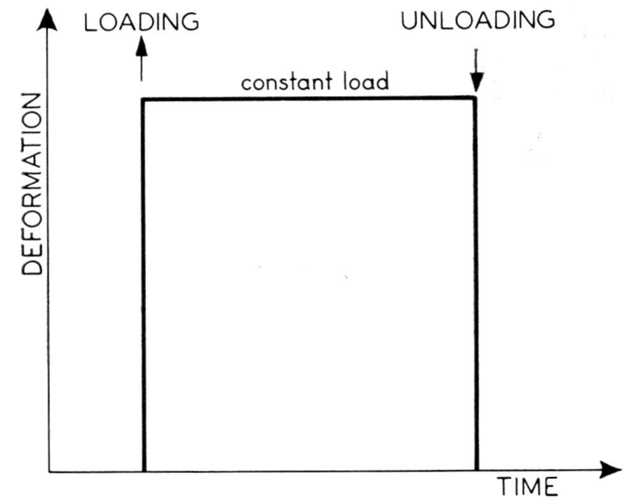

Fig. 4–3. Load-time deformation curve for Hookean body.

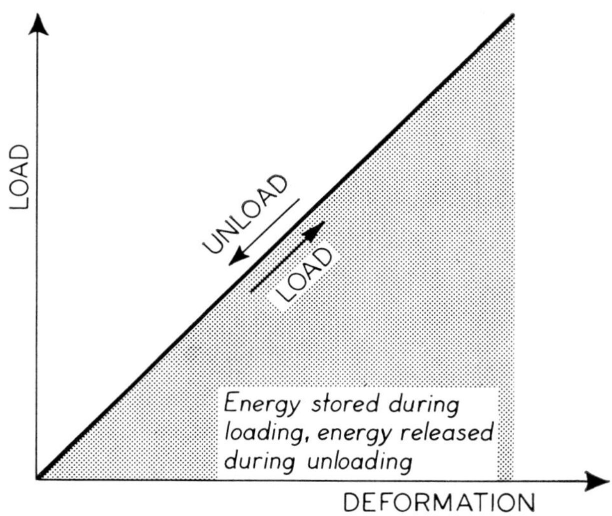

Fig. 4–4. Load-deformation curve for Hookean body. The energy stored during loading and the energy released during unloading are represented by the shaded area under the curve.

Energy is required to deform the Hookean body. The spring has the property of storing the energy in the form of strain energy. The energy input may be calculated by means of the integral of the force multiplied by the displacement. This can be demonstrated graphically as the area underneath the load-deformation curve, as shown in Figure 4–4.

Energy will be released when the spring is unloaded. The energy released can be calculated by measuring the area under the deflection curve. Since the loading and the unloading curves are coincident, the energy stored is equal to the energy released, and there is no net energy loss in the loading cycle.

The other basic element in a simple viscoelastic model is the Newtonian body. This body may be thought of as an idealized dashpot or syringe (Fig. 4–5). The dashpot is a device which uses the viscous properties of a fluid. Because the fluid is viscous, the velocity with which it flows through a given orifice is proportional to the pressure applied to it. When fluid is rapidly injected from a syringe, a much greater force is required on the plunger than when the fluid

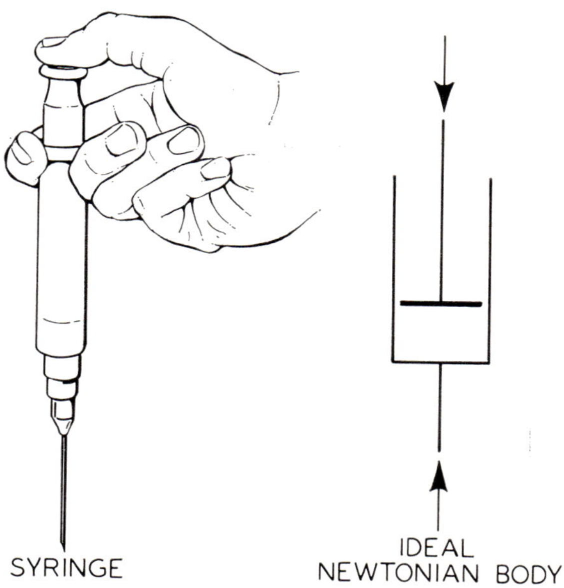

Fig. 4–5. The idealized viscoelastic element or Newtonian body. A syringe is one example.

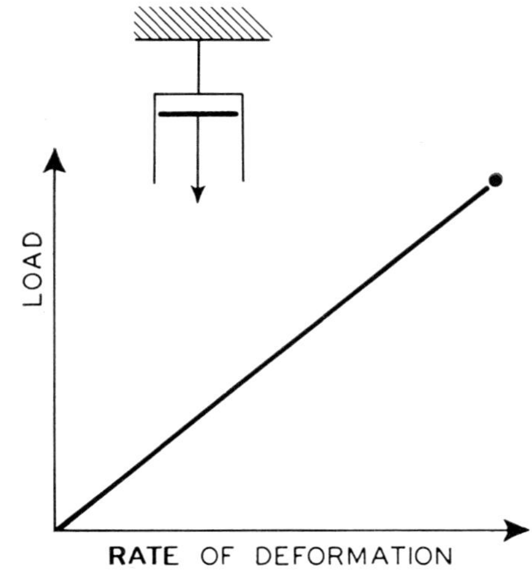

Fig. 4–6. Load versus rate-of-deformation curve for Newtonian element.

is injected very slowly. To achieve the same deformation rate, greater force is required if the coefficient of viscosity is greater, *e.g.,* water and blood.

The behavior of the Newtonian body under load will be drastically different from the behavior of the Hookean body under load. The rate of deformation, not the deformation, of the Newtonian body is a direct function of the load, as shown in Figure 4–6. In the idealized Newtonian body, the relationship between the loads and the velocity (strain rate of deflection) is linear (Fig. 4–7B). The deflection of the dashpot is therefore a function of the time the load is allowed to remain.

As in the Hookean body, energy is required to deform the Newtonian body. The energy may again be calculated as the integral of the product of the force and the distance the force has moved. This can be seen graphically as Area I underneath the load-deformation curve in Figure 4–7A. Unlike the Hookean body, when the Newtonian body is unloaded, there is no tendency to return to its original dimension. No energy is returned when the load is removed. If it is desirable to return the Newtonian body to its original configuration, energy, as shown by Area II, must again be put into the system. The Newtonian body stores no energy. Each change of position of the plunger represents an energy loss. In a complete loading and unloading cycle, energy is lost as opposed to the behavior of an

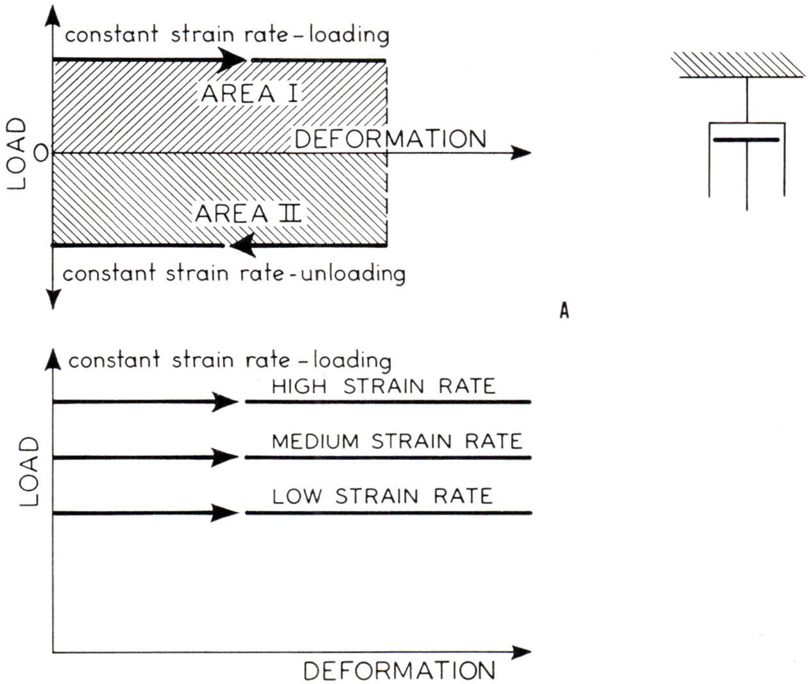

B

Fig. 4–7. **A,** Load-deformation curve with constant strain rate loading of Newtonian body. The energy absorbed by the body during loading is represented by Area I. Area II represents the energy during the unloading cycle. Since Area II is a negative area, that is, it falls below the deformation axis, energy must be put into the system in order to unload it. **B,** The load at which deformation occurs is dependent upon the strain rate which is used to pull the Newtonian body apart.

idealized spring, in which energy is stored as strain energy and then released.

If a spring and dashpot are placed parallel to each other, the resulting combination is called a Kelvin body, as shown in Figure 4–8. The properties of the body may be studied in either of two ways. The deformation under a constantly increasing load may be examined or the load caused by a constantly increasing deformation may be studied. It is important to understand this distinction in developing testing

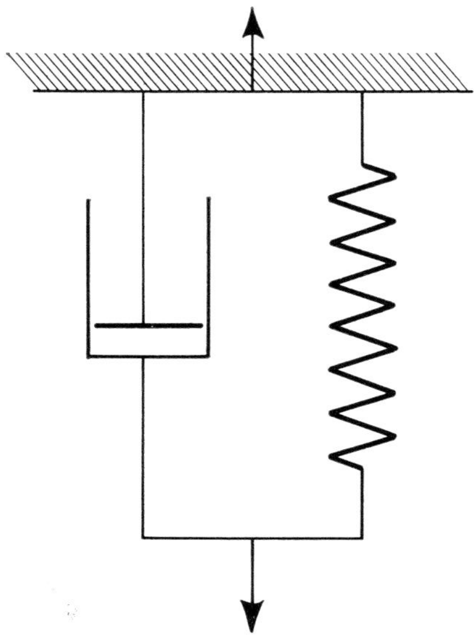

Fig. 4–8. The Kelvin body is the parallel combination of a Newtonian body and a Hookean body.

methods for biological materials since they exhibit viscoelastic behavior. They will vary in their behavior according to the techniques of the investigations. Studies performed under constant load rate may display different behavior than those experiments performed under constant strain rate.

If a constantly increasing load is applied to a Kelvin model the nonlinear load-deformation curve that results is shown in Figure 4–9A. The curve is asymptotic to a straight line whose slope is equal to the spring constant. The load-deformation curve resulting from a constant deformation rate is different from the load-deformation curve produced by a constantly increasing load, as shown in Figure 4–9B. This curve has two straight-line portions, with the slope of the second portion of the curve again equal to the spring constant. The first part of the curve is a reflection of the immediate application of the velocity across the Newtonian body. Unlike the Hookean elements, the Newtonian element generates a resisting force immediately, owing to the applied velocity, even though there is no initial deformation. This behavior is not displayed under the condition of constant loading rate since there is no immediate initial velocity.

If the loading direction is reversed under the conditions of a constant loading rate, a hysteresis loop or energy loss is shown to occur

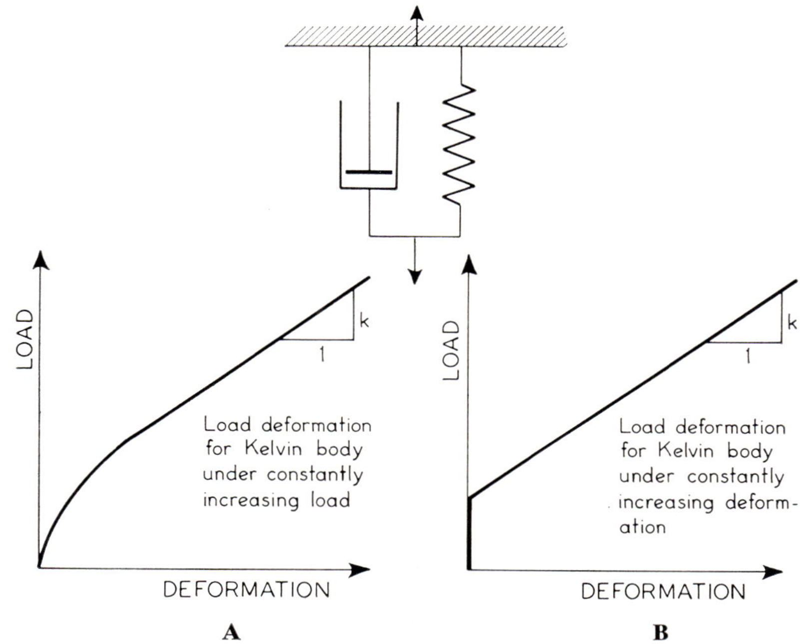

Fig. 4–9. **A,** The load-deformation curve for a Kelvin body produced by a constantly increasing load. The curve is asymptotic to a line of slope k. **B,** Load-deformation characteristics for a Kelvin body under the application of a constant strain rate.

over the duration of the cyle (Fig. 4–10A). A hysteresis loop is also evident in the constant deformation rate test, but displays a different shape (Fig. 4–10B).

When a constant load is suddenly applied to the model, the variation of the deformation produced as a function of time is shown in Figure 4–11. This curve is asymptotic to the value of deformation which is obtained if the load is applied to the Hookean body alone. The Newtonian body serves only to regulate the time of approach to this asymptotic value. When the load is released suddenly, the curve returns to the original length of the model. There is no net change in length over the loading and unloading cycle.

There is another simple combination of the Hookean and Newtonian bodies which is possible, a series combination. This series

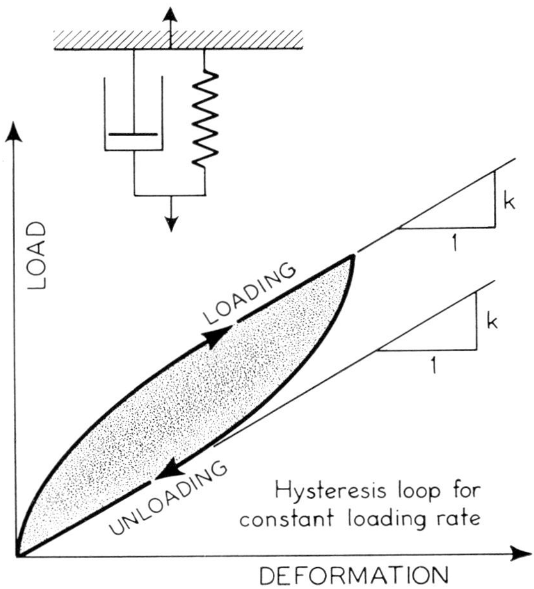

Fig. 4–10A. A hysteresis loop is produced by the Kelvin body when loaded and unloaded at a constant loading rate. The area between the curves represents an energy loss during the loading and unloading cycles.

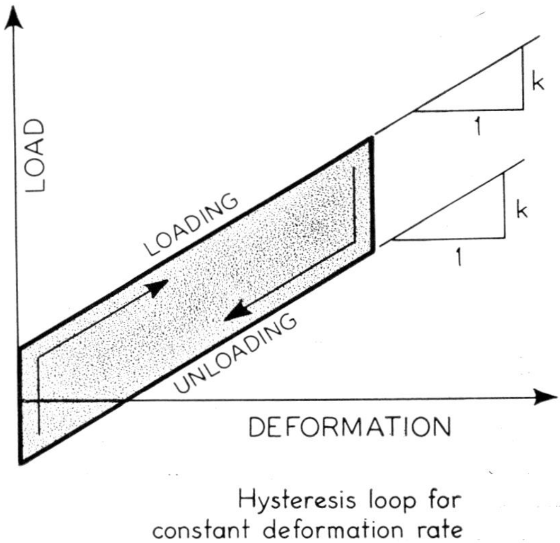

Fig. 4–10B. A hysteresis loop of a different shape is produced if a constant deformation rate is utilized for the loading and unloading cycles.

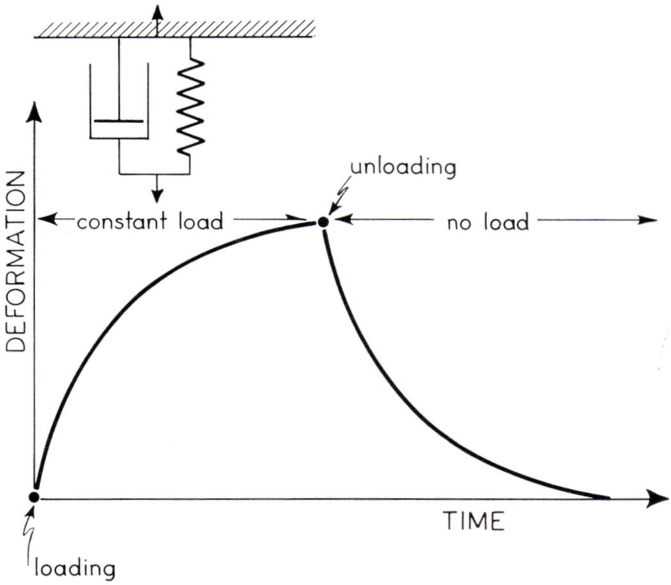

Fig. 4–11. The deformation-time history for a continuously applied load in a Kelvin body.

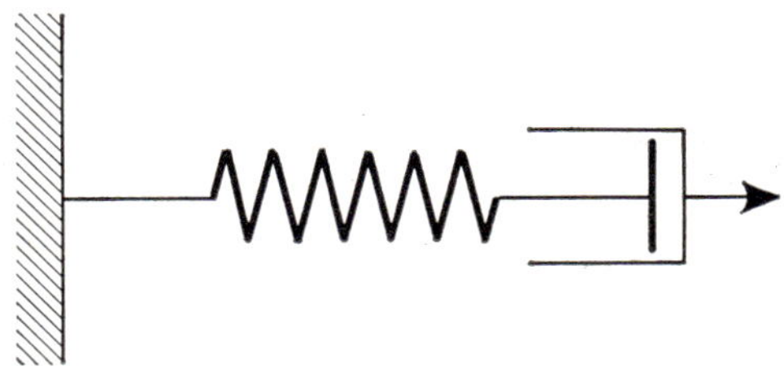

Fig. 4–12. A series combination of a Hookean body and a Newtonian body is known as a Maxwell body.

combination is known as the Maxwell body (Fig. 4–12). Again, the behavior of the body under conditions of constant loading rate and constant deformation rate is markedly different. Under a constant loading rate both load and deformation continue to increase, as shown in Figure 4–13A. For constant deformation rate, however, the load reaches an asymptotic value while the deflection continues to increase (Fig. 4–13B). This asymptotic value is equivalent to the resistance generated by the Newtonian body alone if it is subjected to the same deformation rate. The elongation of the Hookean body is limited to the elongation produced by this asymptotic value of load. The hysteresis effects introduced in each of these loading

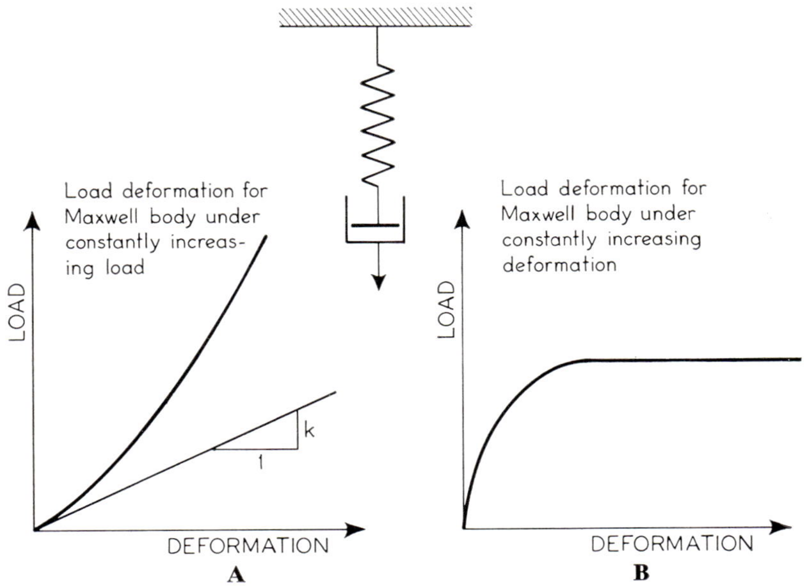

Fig. 4–13. **A,** The load-deformation characteristics for a Maxwell body under a constantly increasing load. **B,** The load-to-deformation characteristics for a Maxwell body under constantly increasing deformation.

procedures is evident in the curves of Figure 4–14. Unless the loading procedure is reversed (*i.e.,* a compressive load applied), the model will not return to its original dimension. In reversing the deformation, any final length may be achieved independently of the original length by varying either the rate of compressive loading or the rate of foreshortening.

Considering the deformations caused by a constant load (rapidly applied) as a function of time, there is an initial rapid deformation due to the elongation of the Hookean elements and then a continued elongation due to the Newtonian elements, as shown in Figure 4–15. Upon release of the load, the system does not return to its original dimension. The body has suffered a permanent deformation which can be reversed only by the reversal of the loading direction. Therefore, while it is possible to create loading cycles which would result in a zero net deformation in general, loading cycles will terminate with some net deformation of the body.

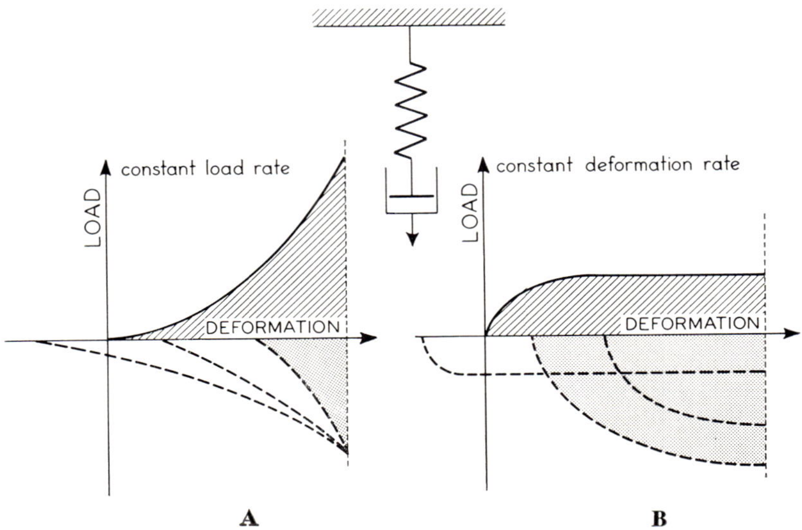

Fig. 4–14. **A,** Possible loading and unloading cycles in a Maxwell body. In order to return a Maxwell body to its original length, the rate of load application in the reverse direction must be carefully controlled. It is possible, by the application of a large loading rate, to return the Maxwell body to a length different from its original resting length. Various loading rates such as 1, 2, and 4 produce different final lengths of the Maxwell body. **B,** Under the constant deformation lengths, various hysteresis curves can be developed, again depending upon the return-deformation rate. Also, the time of application of the return-deformation rate is critical in determining the final point of the length of the Maxwell body. In both A and B, energy is lost both during loading and unloading cycles.

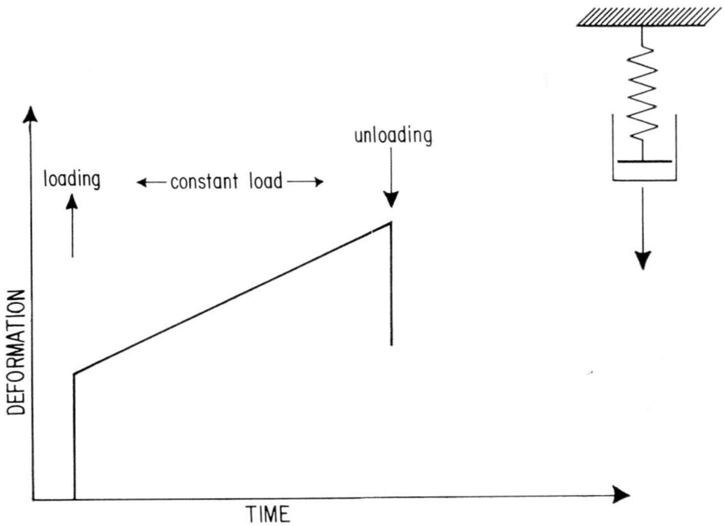

Fig. 4–15. The deformation-time characteristics of a Maxwell body. Upon the application of a load a sudden deformation occurs. Continuation of the load provides a constantly increasing deformation. When the load is removed there is some return to the original position, but it is not complete.

If three elements are allowed in a description of a model, the combination possibilities increase greatly. One of the most useful of these combinations is the addition of a series of Hookean elements to the Kelvin body (Fig. 4–16). Consideration of the load-deformation curves for constant loading rate and constant deformation rate for this type of body shows a less startling difference between the two methods of loading than did a similar examination of the simpler bodies. Both curves are asymptotic to a slope value of:

$$\frac{K_1K_2}{K_1 + K_2}$$

Each curve is different, however, in the rate of approach to the asymptotic value. It should be carefully noted that the curves are not straight lines. Theoretically, no portions of the curves are straight, although as the curves approach their asymptotic value, they do tend toward straight lines. The energy absorbed during a cycle of a

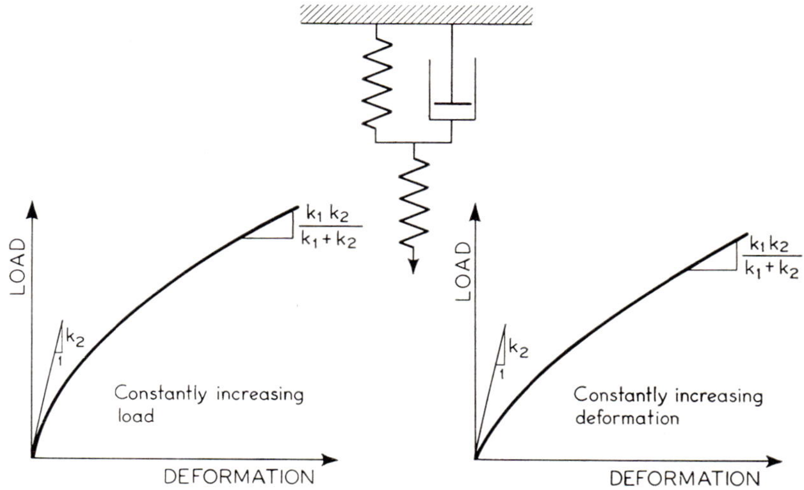

A **B**

Fig. 4–16. A, Behavior of model with constant loading rate. **B,** Behavior of model with constant deformation rate. k_1 = Spring constant in the Kelvin body. k_2 = Spring constant of the series Hookean element.

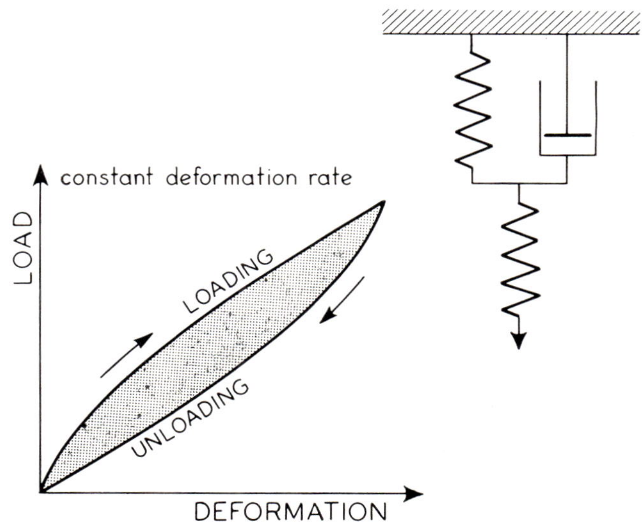

Fig. 4–17. Loading-unloading curve for constant strain rate. The shaded area is a hysteresis loop and represents energy loss during the loading and unloading cycle.

loading and unloading is represented by the area within the hysteresis loop (Fig. 4–17).

A study of the time-dependent deformation caused by a constant load which is rapidly applied is of great interest in this particular model, as shown in Figure 4–18. It can be seen that there is a period of instantaneous displacement followed by a period of exponentially increasing deformation. This increasing deformation again reaches some asymptotic value, which is equal to that deformation which would be suffered by the two Hookean elements if the Newtonian body were absent in the model. In this case, the Newtonian body serves as a retarding element for a portion of the deformation. This behavior very closely approximates the behavior of the earlobe tissues described earlier in Figure 4–1. The unloading process in this cycle is characterized by initial, instantaneous recoil toward the original dimension, followed by an exponential decrease which is asymptotic to the original position. When the load is removed, the body returns to its original shape after a sufficient lapse of time. In many materials the amount of elongation which occurs at constant load is many times the amount of elongation which occurs over the initial elastic region.

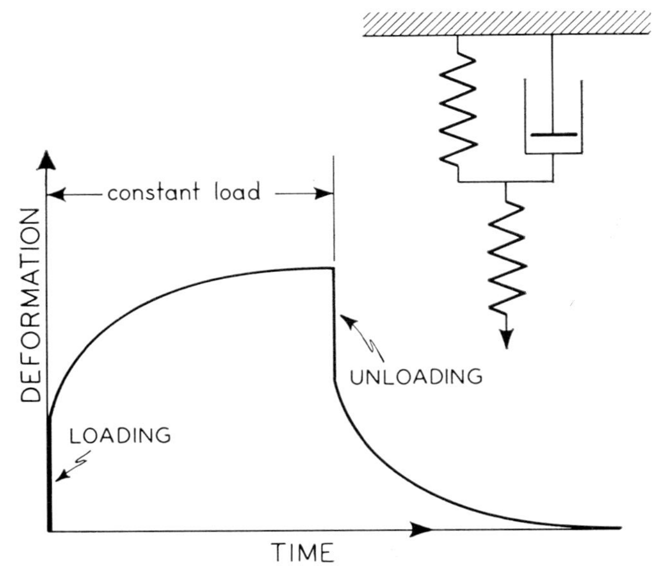

Fig. 4–18. Time-deformation characteristics of combination body for a loading and unloading cycle utilizing constant load.

The constant load elongation or viscoelastic flow, as it is sometimes called, may comprise a significant portion of the energy-absorbing capabilities of a material.

The application of the term "modulus of elasticity" to load-deformation curves of viscoelastic materials is, at best, quite arbitrary. It is evident that the modulus of elasticity, which is actually a measure of the slope of such curves, is constantly changing. If the asymptotic value of the curve is chosen as a value of the modulus of elasticity, it must be recognized that this value may be inaccurate as a representation of material behavior in a region of small deformation.

The value of a theoretical model is realized when prediction based on the behavior of the model is matched by the observed behavior of the biological tissue that it is intended to simulate. Under conditions of a rapidly applied constant load, the ideal model, which is composed of a parallel Hookean and Newtonian body in series with a Hookean body, displays a deformation-time characteristic markedly similar to that of biological tissue. It is, therefore, desirable to predict the response of bony tissue to the varying strain rates on the basis of

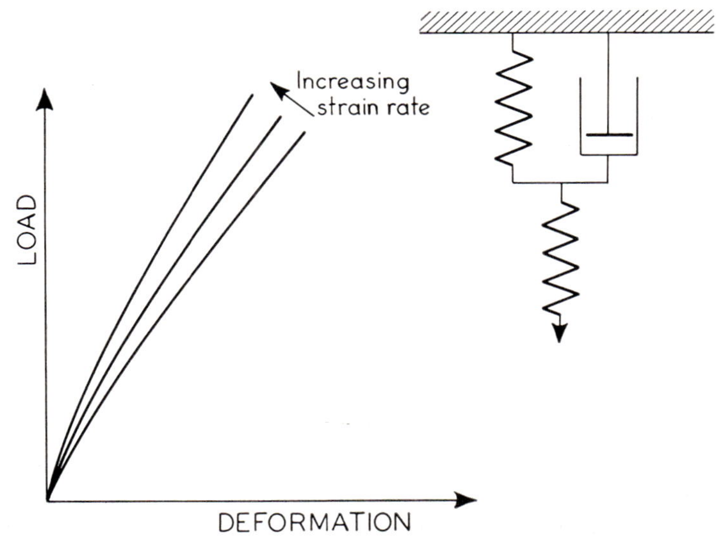

Fig. 4–19. The predicted behavior of a model comprised of a parallel Hookean and Newtonian body in series with a second Hookean body. The effects on the load-deformation characteristics of increasing strain rate are demonstrated.

the mathematical model. The mathematical response of the model shows that, at increasing strain rates, the material should respond with increasingly greater initial slopes (Fig. 4–19). The theoretical response produces a constant value of slope after the transient response due to the dashpot's diminishing. It must be remembered that these theoretical load-deformation curves are shown with arbitrary termination points since the simple theoretical model which is used here in no way takes into account any condition of yield or failure. Thus, no limiting values of elongation or limiting values of stress have been inserted into the model behavior. This is an important consideration when the response of the idealized model is compared to the response obtained from biological tissues. The results of tests conducted of varying strain rates on bovine bones are shown in Figure 4–20. There is excellent qualitative agreement, up to the point of yield, between the model behavior (Fig. 4–19) and the experimentally observed behavior. Several important concepts can be obtained from a close scrutiny of both the theoretical model and the results

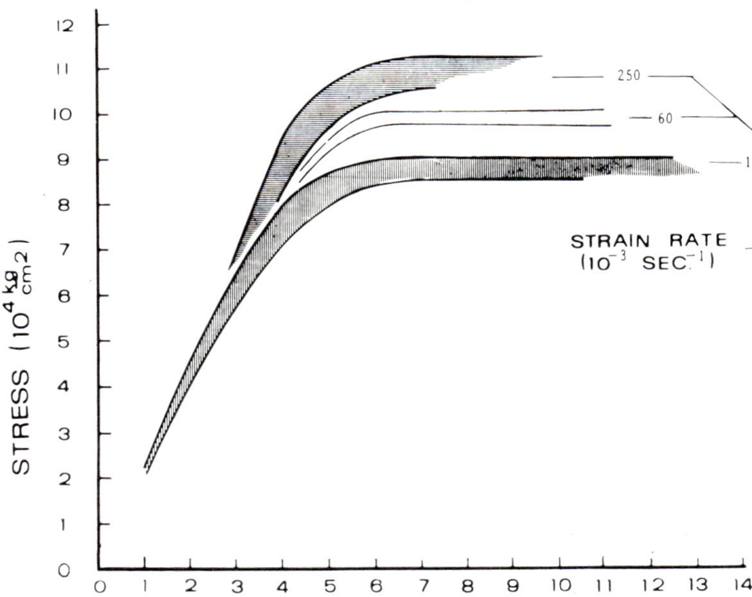

Fig. 4–20. Load-deformation curves for bovine bone loaded at varying strain rates. The upper portion of the curve represents a definite yield area.

obtained from biological tissues. It is apparent in the theoretical model that the amount of energy absorbed up to the point of failure depends upon the failure criteria. If we arbitrarily choose a failure criterion which states that failure occurs when a predetermined amount of elongation of the specimen is encountered, those specimens which are loaded at the more rapid strain rates will obviously be able to absorb more energy (Fig. 4–21A). Investigation of the area underneath the load-deformation curve confirms this. Again it must be stated that the mode of material failure is an arbitrary assumption. Had the assumption been made that the cause of failure was the attainment of some arbitrary limiting value of stress, the specimens

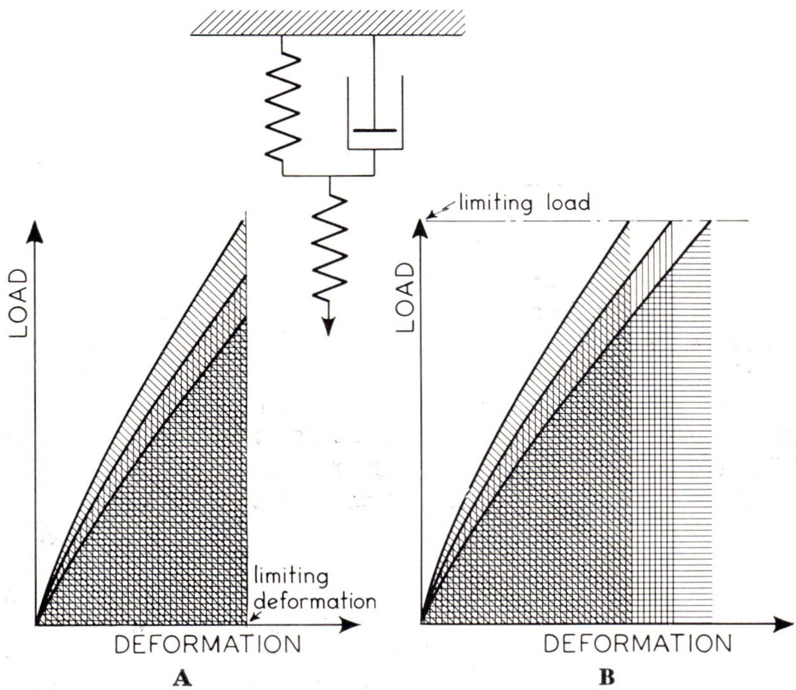

Fig. 4–21. **A,** The amount of energy absorbed at varying strain rates up to the point of failure varies with the strain rate. In the theoretical model, if a limiting deformation is chosen as a failure criterion, the higher strain rates result in the absorption of more energy. **B,** If a limiting value of load is chosen as the failure criterion, the lower strain rate allows the model to absorb more energy.

loaded at the lowest strain rates would have absorbed the greatest amount of energy to failure (Fig. 4–21B). Again, an examination of the areas underneath the load-deformation curves confirms this. The criteria for failure may be either a combination of the two criteria chosen above, one of the two criteria individually, or some more complex combination of deformation, load, and perhaps energy level. This is the case in the bovine samples, in which yielding occurred before failure. It is entirely possible that the criterion is not a constant one; that is, the criteria for failure may be dependent upon the rate of loading.

The effects on energy absorption caused by the viscoelastic properties of bony material may be summarized in the following manner: Each loading and unloading cycle which the tissue encounters will cause a retention of energy within the tissue. This is because all loading cycles are of finite duration and the bony material displays a hysteresis loop when subjected to a loading and unloading cycle. In addition, the term "modulus of elasticity" and its associated term "modulus of rigidity" can no longer be thought of as constants or as inherent properties of the material unless they are redefined so as to include an identification of the time-dependent loading conditions under which the tests are made. A higher strain rate characteristically produces higher values of the "modulus of elasticity." The viscoelastic property of biological tissue generally allows the tissue to absorb different amounts of energy when loaded at differing rates. Whether the energy capacity before failure is increased or decreased with an increasing loading rate can be determined only when the failure criteria are identified. The absorption of energy into a material usually results in a temperature rise within that material.

Since the concept of yield is important to the understanding of the physical behavior of biological materials, it is of significant importance to be able to include in the theoretical model a mechanism which allows yielding. Such a mechanism is the St. Venant body, which is characterized as an ideal friction element. The load-deformation of such a body is shown in Figure 4–22. The deformation which occurs across such a body occurs at constant load. Until a critical value of load is reached the body undergoes no deformation. Once this critical value of load is obtained, the body is free to move so long as the applied force moves with it. This body displays no mechanism which would allow it to resist a force greater than that required to move the body initially. Upon release of the applied

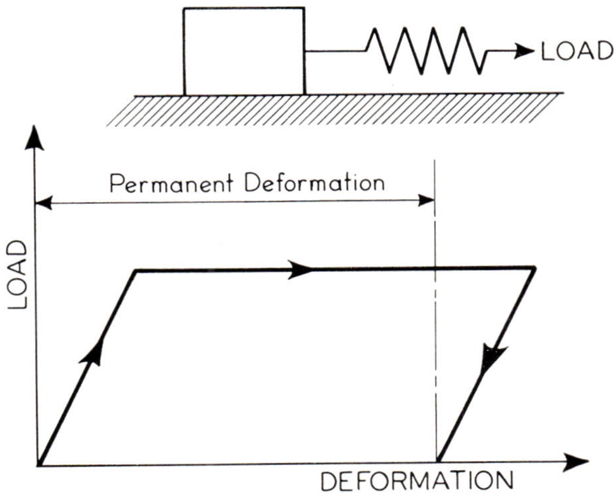

Fig. 4–22. The load-to-deformation characteristics of a St. Venant body. Until a critical value of load is obtained, no deformation occurs. Once this load is obtained, however, deformation may occur freely without any increase in applied load.

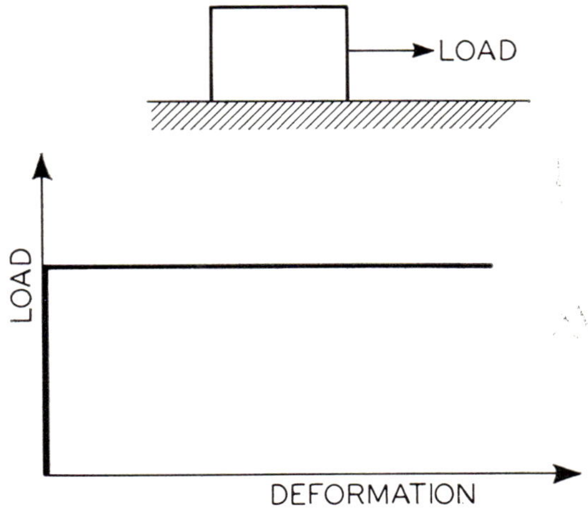

Fig. 4–23. The load-deformation behavior of the viscoelastic model with the addition of a St. Venant body. The St. Venant body allows permanent deformations to occur when sufficiently high loads are reached. Compare with Figure 4–20.

load, the body remains stationary. It has no tendency to return toward any original position. Whatever energy is expended in the elongation of such an idealized body is irretrievably lost into the body. The body is not sensitive to loading rates or strain rates. The total deformation suffered by the body, however, is a function of the duration through which the load is applied.

This type of idealized body is useful in the description of yielding phenomena and can best be illustrated by considering the series combination of a Hookean body with the St. Venant body. The load-deformation characteristics of such a combined body are shown in Figure 4–23. As the load is applied, the Hookean element elongates in the usual linear elastic manner. Once a critical value of elongation has been reached, the Hookean body develops a force sufficient to cause the deformation of the friction body. When this condition has been realized, the magnitude of the applied force cannot be increased. The sustained application of this force only causes further deformation within the body. Upon release of the applied load, the load-deformation curve returns to the base line, as shown in Figure 4–23. The body thus sustains a permanent plastic deformation.

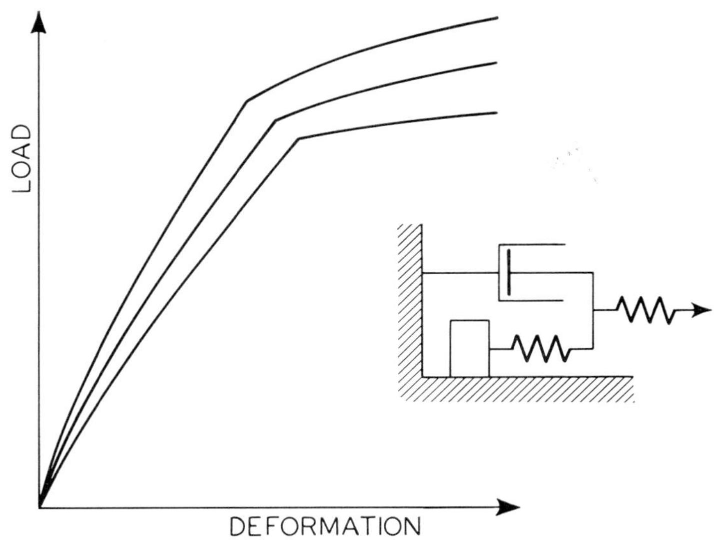

Fig. 4–24. Plastic strain can be included in a model if a St. Venant body is introduced. The value of the load at a particular deformation increases with increasing strain rate.

The St. Venant body may be added to the model for biological tissue in several ways. One such way is considered here. Figure 4–24 illustrates the load-deformation curve produced by the model. The permanent deformation exhibited by the model is a function of the strain rate as well as of the load.

Biological tissues exhibit viscoelastic behavior to varying degrees. Studies of the mechanical behavior of living tissues must take into account this complex phenomenon.

5. Kinematics

KINEMATICS is the study of the relative motion that can exist between rigid bodies, which are called links. Although no body is absolutely rigid, the deformations which are induced in biological links such as bone are very small compared to the gross motion of the bone. That is, when the knee joint is extended from a flexed position, the tibia undergoes a gross, visible angular displacement on the femur. Simultaneously, the forces which have caused the displacement have also produced an internal effect in the femur and tibia. The internal effect of a force system as noted in Chapters 2, 3, and 4 is to cause a state of strain. Though measurable with suitable strain-detecting instrumentation, the changes in length are negligible when compared to the gross displacement of the tibia on the femur. For the purposes of kinematic study, the bone can be considered to be a rigid structure.

In studying kinematics, only the gross relative motion between bony structures is considered. Any bone in the body is considered to be a rigid body which possesses certain definite characteristics. The first and most important of these characteristics is the fact that any two points on a rigid body or link can never approach each other or recede from each other. This is another way of stating that the distance between two points on a rigid body can never change. The second property of a rigid body is that its boundaries may be considered extensible. For purposes of analysis the dimension of the rigid body may be considered as being greater than that of the actual bone which it represents. Although a bone such as a femur has a well-known physical shape, for purposes of kinematic analysis its boundaries may be expanded to any hypothetical shape. It is reasonable to represent a point A as being a point on rigid body 1 even though point A does not actually lie within the physical bone which is represented by body 1 (Fig. 5–1). Each point on the rigid body may, over an interval of time, undergo a linear displacement. Displacements are simply the change in the position in space of a point on a rigid body. It must be noted that the displacement is a vector quantity having direction, magnitude, and sense (Fig. 5–2). In addition, the rigid body itself may undergo an angular displacement during a period of time. An angular displacement is a change in angular

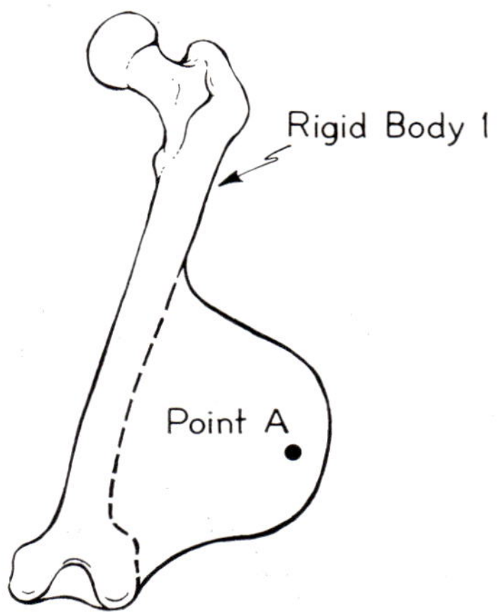

Fig. 5–1. Point A may be considered as part of the link (Rigid Body 1) even though it does not physically lie on it.

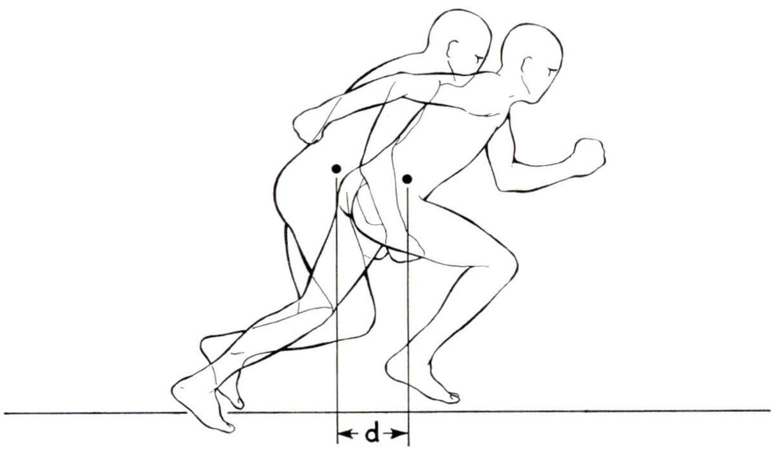

Fig. 5–2. Linear displacement is a change in linear position of a particle (d).

position of any straight line on the rigid body. The angular position is measured in radians (Fig. 5–3). All lines on the rigid body undergo equal angular displacements in equal time periods. This is not true of the linear displacements of points on the rigid body. This is seen in Figure 5–2. The angular displacements do not depend upon any particular center of rotation. As shown in Figure 5–3, when the rigid body moves from position 1 to position 2, the angular displacement of the body is angle β-α.

The concept of velocity is based upon the concept of vector displacements. The linear velocity of a point on a link is simply the ratio of the linear displacement of that point to the amount of time required to obtain that displacement. If the time period is of finite duration, for example $1/10$ of a second, then the average velocity obtained by dividing the displacement by the time of $1/10$ of a second is the average velocity over the time period. The concept of instantaneous velocity differs slightly in that the value of the time period over which the displacement occurs is made extremely small. Instantaneous velocity, then, is the rate of change of position with respect to time.

Velocity is a vector quantity. It possesses magnitude, direction, sense, and point of application. The magnitude of the velocity is often called speed and is measured in such units as feet per second or miles per hour. Direction is the line along which the velocity is acting. During the gait cycle, the velocity of the foot may be horizontal. The sense of the velocity describes whether the foot is going forward or backward. Point of application refers to the particle whose

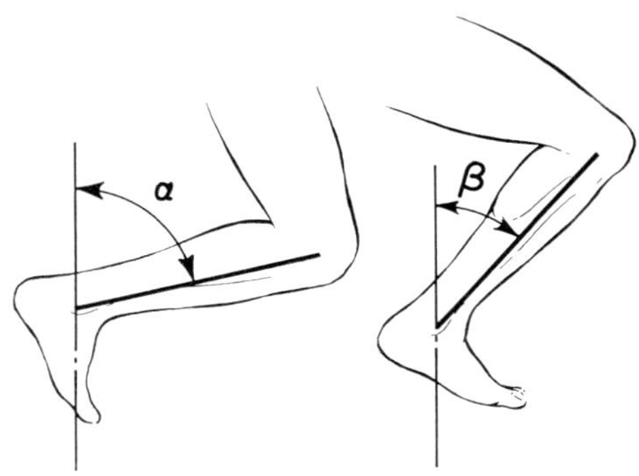

Fig. 5–3. Angular displacement is a change in angular position of a line (α–β).

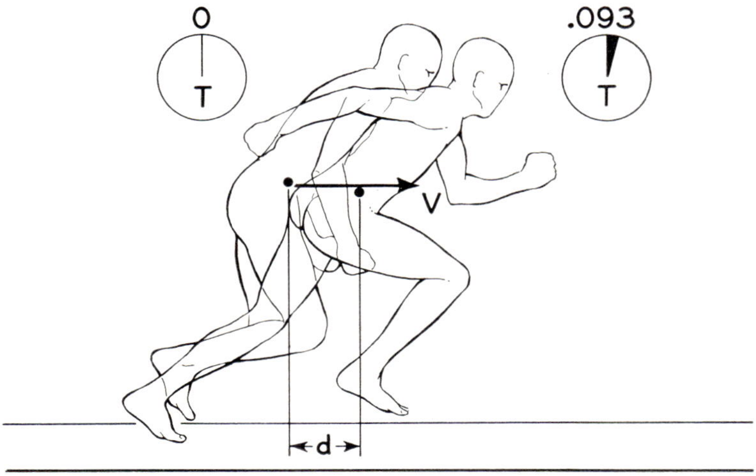

Fig. 5–4. Linear velocity is the time rate of change of the linear position of a particle, $\frac{d}{T}$.

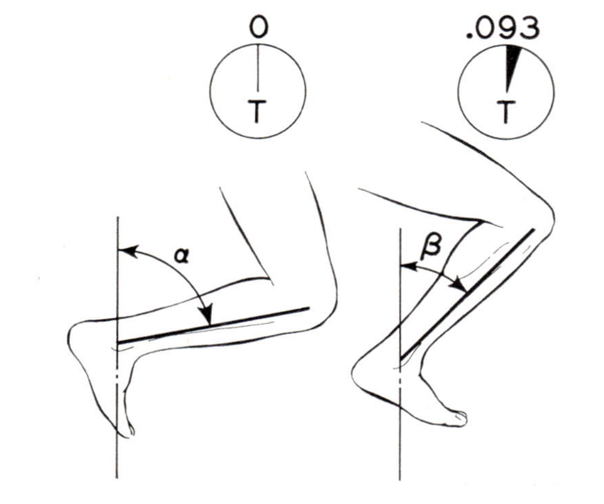

Fig. 5–5. Angular velocity is the time rate of change of the angular position of a line, $\frac{\alpha-\beta}{T}$.

velocity is being described (Fig. 5–4). A similar velocity concept can be deduced from angular displacement (Fig. 5–5). An average angular velocity is obtained if the angular displacement is divided by the time duration over which the angular displacement occurs. If the time duration is allowed to become very small, the ratio of the angular displacement to the time duration is the instantaneous angular velocity. Again, the time duration must be made to approach zero as the limit for this to become strictly true.

The concept of linear acceleration is an extension of the previous ideas (Fig. 5–6). Linear acceleration is a measure of the rate of change of linear velocity. If the velocity change occurring over a time interval is divided by the magnitude of the time interval, the result is the average acceleration over the time interval. If the time duration is allowed to become very small, the resulting acceleration is the instantaneous acceleration of the point. Acceleration, therefore, is a measure of the time rate of change of velocity. Acceleration is also a vector quantity. As such, it possesses magnitude, direction, sense, and point of application in a manner similar to force, displacement, and velocity.

This same concept may be applied to angular motion. The result

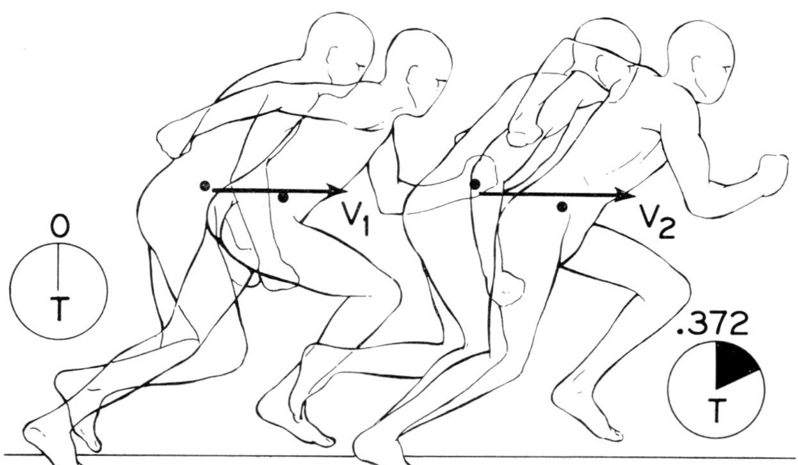

Fig. 5–6. Linear acceleration is the time rate of change of the linear velocity, $\frac{V_2 - V_1}{T}$.

is that angular acceleration is the time rate of change of angular velocity (Fig. 5–7).

It is also possible to measure the time rate of change in acceleration in a similar manner. This rate of change is termed jerk.

In Figure 5–8 a small body which is rotating about a fixed center is shown in three positions with respect to time. The particle is considered to be attached to the center of rotation by means of a string and is rotating in a horizontal plane. In going from position 1 to position 2 the string undergoes an angular displacement (α). At the same time the particle undergoes a linear displacement (d). If position 1 and position 2 are sufficiently close together, the linear displacement (d) coincides with a small portion of the circumference of the path. If the time duration between these two successive positions is infinitesimally small, the instantaneous velocity of the particle will lie in the same direction as the displacement and will have the same sense. Note that the direction of this velocity is tangential to the path. If the particle is observed during the next time duration, the direction of the velocity is seen to change, and since the change in any vector quantity implies a change in the vector, the velocity itself changes from position to position. Note that the magnitude of the velocity, or speed, has remained the same.

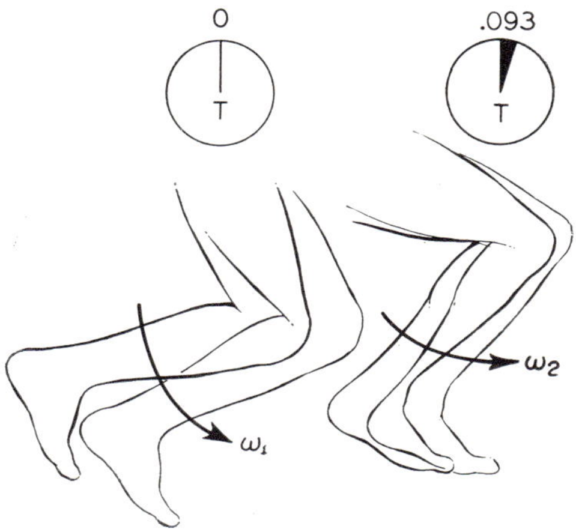

Fig. 5–7. Angular acceleration is the time rate of change of the angular velocity, $\frac{\omega_2 - \omega_1}{T}$.

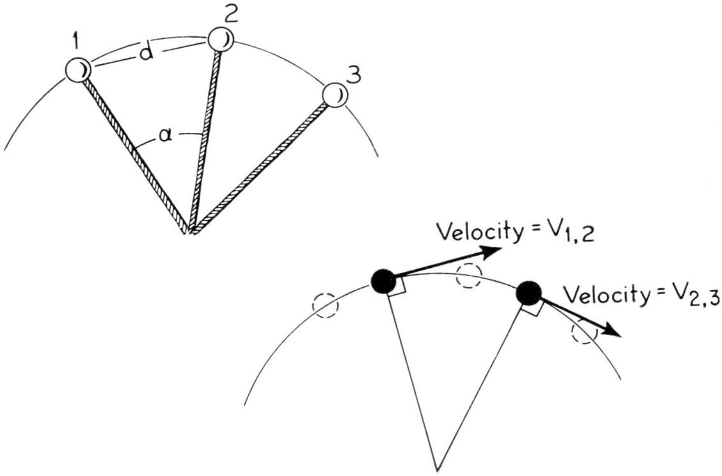

Fig. 5–8. The average velocity between positions 1 and 2 is $\mathbf{V}_{1,2}$. Between positions 2 and 3 the average velocity is $\mathbf{V}_{2,3}$.

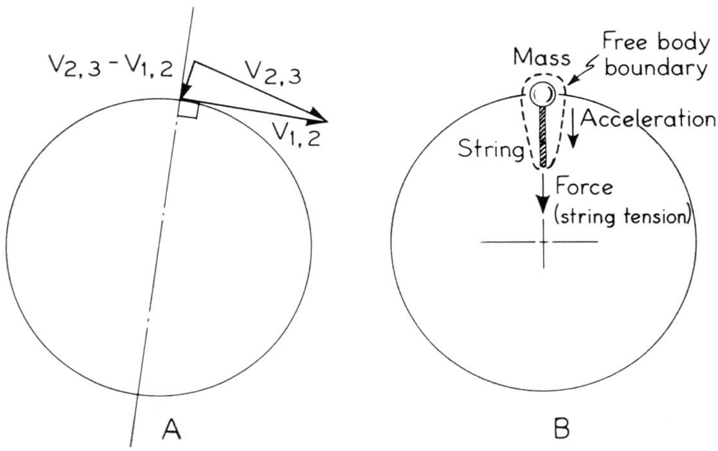

Fig. 5–9. A, The change in velocity ($V_{2,3}$–$V_{1,2}$) is seen to point to the center of the circle. **B,** This is the direction of the acceleration of the mass.

Since the direction of the velocity and thus the velocity have changed, the particle has undergone an acceleration. Any change in the magnitude, sense, or direction of a velocity indicates the occurrence of an acceleration.

The change in the velocity of the particle is simply the difference between the two velocities. Vector subtraction is shown in Figure 5–9A. The difference in the two velocities of $V_{2,3}$–$V_{2,1}$ is seen to point in the direction of the center of the circle. Since the acceleration is the difference in the velocities divided by the elapsed time, the direction of the acceleration must be along a radial line, with sense towards the center of the circle.

From Chapter 1 it will be recalled that in order to have a body undergo an acceleration, a force must be applied. In this problem the force which is causing the radial acceleration is produced by the tension in the string. A free body of the particle with its attached string is shown in Figure 5–9B.

In Chapter 1, the concept of treating the component of a force vector was discussed. A similar concept can be applied to determine the components of the velocity of a point on a body. In Figure 5–10, point A has a velocity V in the direction shown. It may be desirable to determine the component of the velocity of point A in the direction X. This is accomplished by dropping a perpendicular from the arrowhead of velocity V to the line in the X direction. The length V_x is

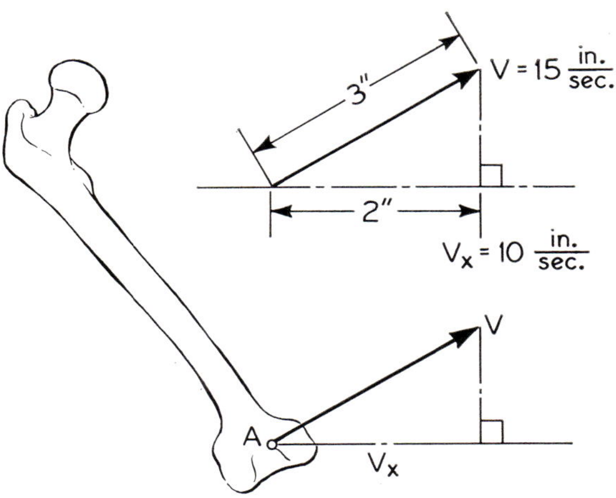

Fig. 5–10. The component of V in the x direction (V_x) is determined by dropping a perpendicular from V to the x direction.

the component of the velocity of point A in the X direction. If the velocity of A is 15 inches per second, the component V_x can be determined by scaling its length and comparing it to the original length of V_x. If V is 3 inches long, each inch represents 5 inches per second. If V_x is measured and found to be 2 inches long, V_x is therefore 10 inches per second.

The velocity component exerts a strong influence on the allowable velocity configuration in a rigid body. Consider Figure 5–11, in which point A has a velocity as shown. Point B is another point on the same rigid body. If a line is drawn between A and B, the component of the velocity of point A, which is in the direction of this line between A and B, can be determined. The component is obtained by dropping a line from the arrow of the vector V_A perpendicular to the direction AB. The component of the velocity of point A in the direction AB is given by vector V_A/AB. Since point B and point A lie on the same link, the distance between these two points cannot change. Therefore, the component of the velocity of point B in the direction AB must be identical to this given component of the velocity of point A in the direction AB. If the two velocity components in the direction AB are not identical, the points either move closer to-

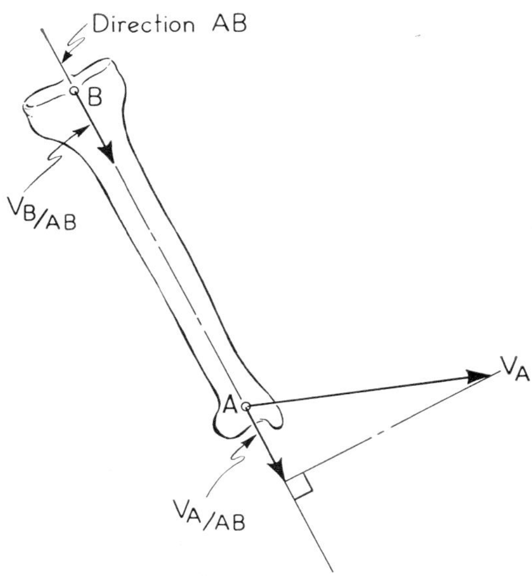

Fig. 5–11. The component $V_B/_{AB}$ must equal $V_A/_{AB}$.

gether or separate. This, for the definition of a rigid body, is impossible. Therefore, there is a restriction on the velocity of B once the velocity of A is determined or given. What then are the permissible values of the velocity of B? B may have any velocity which has as its component in the AB direction the value required for V_B/AB, namely V_A/AB. All velocities which possess this characteristic must lie on a line perpendicular to the direction AB and intersecting AB at the arrowhead, as shown in Figure 5–12. Any of the velocities V_{B1}, V_{B2}, or V_{B3} along with an infinite number of similar velocities are acceptable solutions to the velocity of point B. Notice that all of them have the required components along the direction AB.

Consider now the case of a kinematic body whose motion is prescribed. In particular, examine two points on that body, A and B. Since the motion of the body is prescribed, the values of the velocities of these two points at any instant in time are known. In Figure 5–13, the velocity of A and the velocity of B are given. Consider now a point C on the same body which is located by the intersection of two lines, as shown in Figure 5–13. One line passes through point A and is perpendicular to the velocity of A. The second line passes through point B and is perpendicular to the velocity of B. The velocity of the point C on the given body can now be determined. When the component of the velocity of point A is determined in the direction AC, it is found to be zero. This is because a line drawn from the arrowhead of the vector V_A perpendicular to the line AC intersects

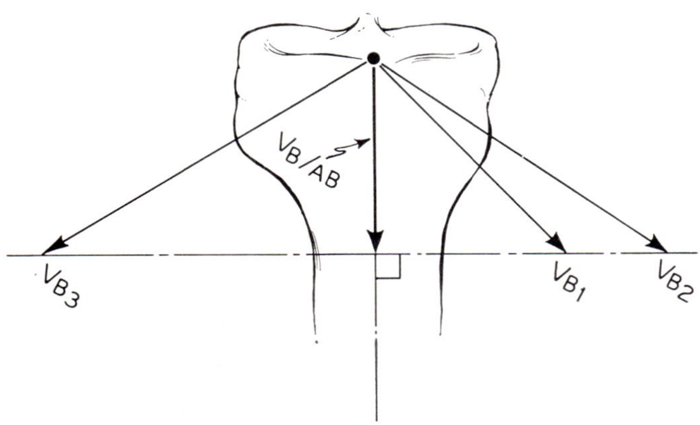

Fig. 5–12. Any velocity such as V_B, which touches the line perpendicular to $V_B/_{AB}$, is an acceptable velocity.

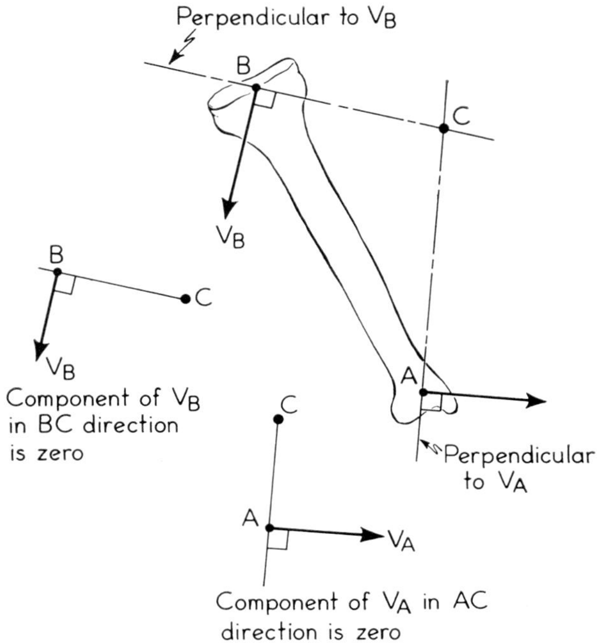

Fig. 5–13. Two components of the velocity of point c are zero.

the line AC at the point A. The vector which represents the components of the velocity of A in the direction AC has zero length and therefore is a zero velocity. The same statements hold if the components of the velocity of B in the direction BC are to be determined. This velocity, too, is zero. Attention may now be directed to point C; its velocity may be determined. The component of the velocity of C in the direction AC must be the same as the component of the velocity of A in the direction AC. This was shown to be zero; therefore, the component of the velocity of C in the direction AC must also be zero. Again, a similar argument can be applied to determine the velocity component of C in the direction BC. It will be seen to be zero also. The velocity of C has the properties of having zero components in the directions AC and BC. The only velocity which can satisfy these two requirements is a zero velocity. Thus point C is seen to be stationary.

Point C is called an instant center of rotation of the body. It may

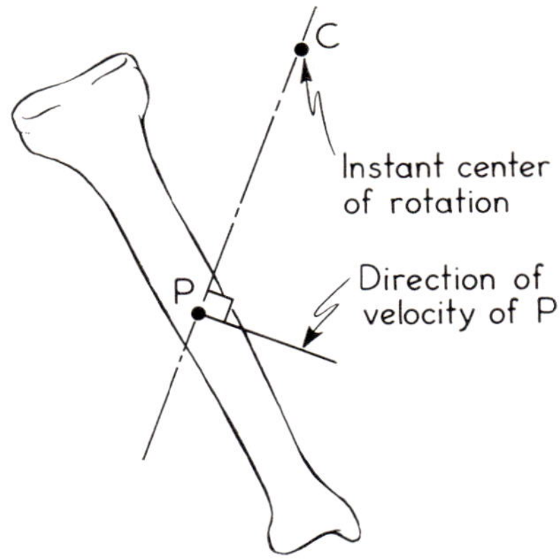

Fig. 5–14. The direction of the velocity of point P is perpendicular to the line between the instant center C and P.

be thought of as a pivot point for this instant in time. The direction of the velocity of all other points in the body may be determined, as in Figure 5– 14, simply by drawing a line from a point, for example, point P to point C, the instant center. The direction of the velocity of P must be perpendicular to the line PC. This is because C is actually the center of rotation of the body at this particular instant and the velocity of any point on a moving body is perpendicular to the direction from the point to the center of rotation.

As the body moves through space, the velocities and, in particular, the direction of the velocities of points A and B change. In the example in Figure 5–15A, points A′ and B′ have velocities V_A and V_B. The body is rotating about the points C_1 and C_1', which are components. The body is moving in reference to the fixed line at the bottom of Figure 5–15A. At some further instant in time, points A_1 and B_1, now at A_2 and B_2, have the velocities as shown in Figure 5–15B. The new instant center of rotation, C′, is shown. C_1 and C_2 do not coincide. In general, for each position of the body in space there will be a new position for the instant center.

The point of instant center may actually be considered as two coincident points. One of the two points may be thought of as fixed to the reference or ground system and the second of the two points may be thought of as lying on the moving body. When the link moves from position one to position two, the two coincident points C_1 and C_1' separate. In the second position, point C (fixed to ground) is shown separated from point C_1'. In Figure 5–15B the instant center has been calculated for the second position of the moving kinematic body. Similarly, points may be obtained for additional positions. Each of these positions has a corresponding instant center which may be considered as either fixed to the reference or moving with the body. In Figure 5–15C, two curves are formed by the instant center points. One curve is depicted as the path of the instant center on the ground or reference body and the other curve is depicted as the curve of the instant centers on the moving body. These curves are called polodes

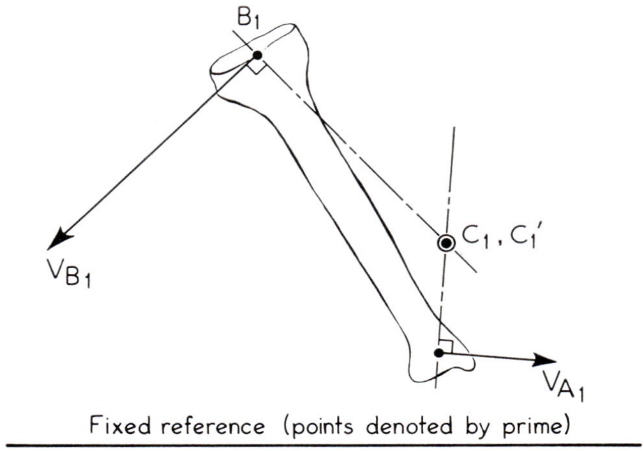

Fig. 15A

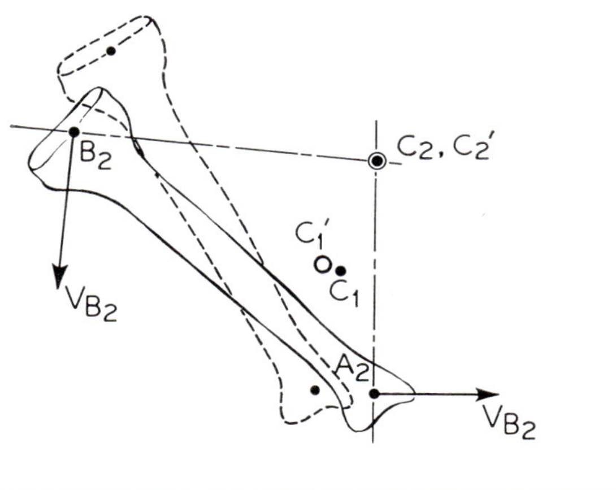

Fig. 15B

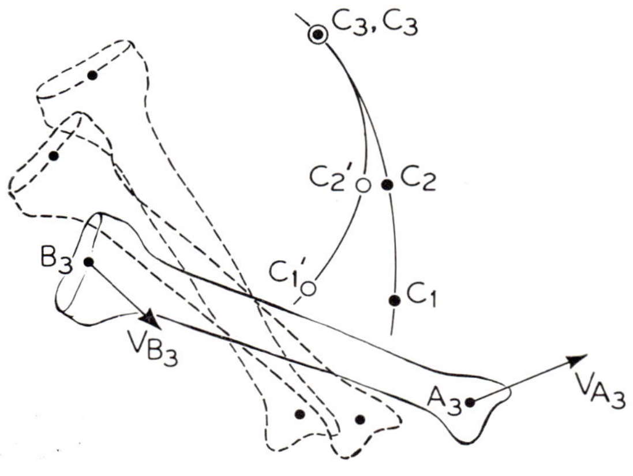

Fig. 15C

Fig. 5–15A, B, and **C.** The tibia moves from position 1 to 2 to 3. The instant centers for each motion are shown. The instant centers with respect to the fixed reference plane (denoted by primes) do not move.

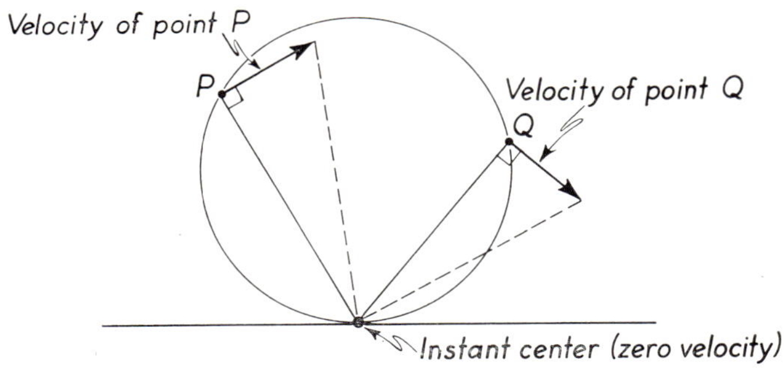

Fig. 5–16. If the instant center should fall on the point of contact between two bodies, that contact point has zero velocity. The bodies are said to be in rolling contact.

or centrodes. The polode curves contain most of the information necessary to visualize or analyze the kinematic motion of a body. In Figure 5–16, the instant center lies on the point of contact of the two bodies. The resulting motion is called rolling, which is defined as the condition that exists when the instant center of rotation between two bodies lies on their point of contact. Note that there is no relative velocity between the two bodies at the point of contact.

The concept of instant centers may be regarded from another point of view (Fig. 5–17). Consider points A_1 and B_1 as belonging to the same link. If the link moves from position (1) to position (1′), point A moves from position A_1 to position A_1'; point B moves from point B_1 to position B_1'. The motion of point A may be considered to be a small portion of an arc; likewise the motion of point B may be considered as a portion of an arc. The center of the arc which will move point A from 1 to position 1′ must lie along a line perpendicular to the line connecting A in its two positions. The same must be true of point B. Its arc must have a center on a line which is the perpendicular bisector of line B_1, B_1'. The only point which will satisfy both these conditions simultaneously is the point of intersection of the two perpendicular lines. This point of intersection is called the pole or the instant center of displacement. The instant center for velocity will coincide with the pole of displacement if the displacement is taken over a very short interval of time. That is to say, the velocity vector can be considered to be overlying the displacement vector if the latter is small enough. By the method of taking a series of successive positions of a link in space the path that the centrode follows may be determined (Fig. 5–18). This path will exactly coincide with the path of the instant centers as described in the previous method.

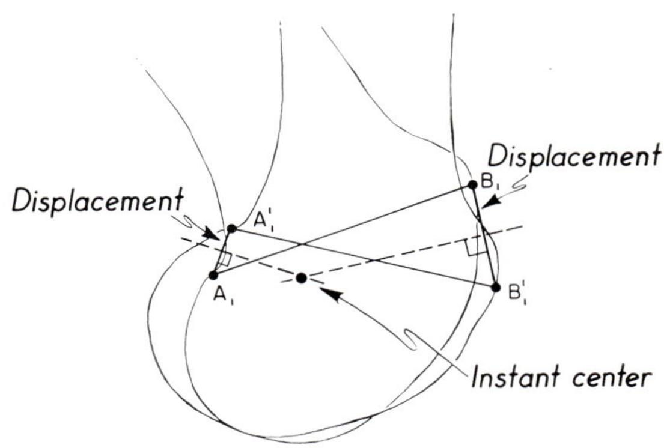

Fig. 5–17. Another method of locating the instant center of a link is to draw perpendicular bisectors of the displacements of the two points on the link.

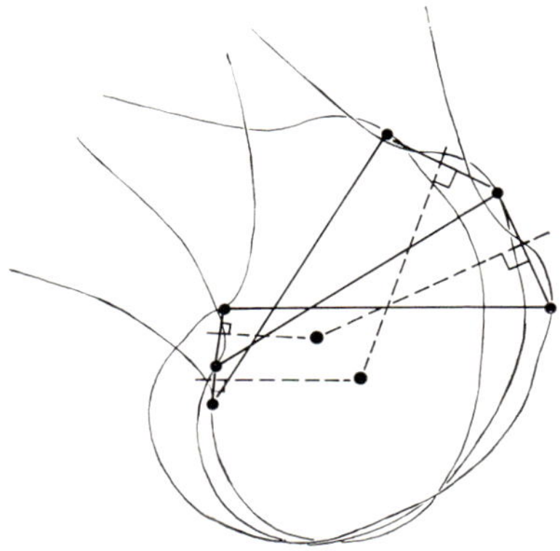

Fig. 5–18. The path of the instant center can also be obtained by knowing the successive positions of the link.

To illustrate the preceding discussion of kinematic principles, a human knee joint will be analyzed and its kinematic function described. Figure 5–19 shows a series of x-ray films in which the relative positions of the tibia, the fibula, and the femur are changed by intervals of approximately 10 degrees. Since kinematic analysis is a study of the relative motion between two bodies, one of the bones will be chosen as the reference system and the other bones will be considered to be rotating about the reference bone. In this case the tibia is the reference. Figure 5–20 shows the superimposition of the x-ray reproductions with the tibia held in relatively the same position while the femur is shown in a relatively varying position. This step of the analysis is quite arbitrary. The results which follow would be identical if the reference body were chosen as the femur instead of the tibia. Points are now identified on the femur and located on each successive position. The vector displacements may be obtained by connecting successive positions of the identified points (Fig. 5–21A). It should be noted that there is an inherent error involved in this procedure since the motion of the knee joint does not take place entirely within the anteroposterior plane. Because a limited amount of rotation occurs during the last phase of extension, the distance between the two chosen points either lengthens or foreshortens slightly when viewed on the plane of projection of the x-ray film. The displacements that are shown are actually the projections of the real displace-

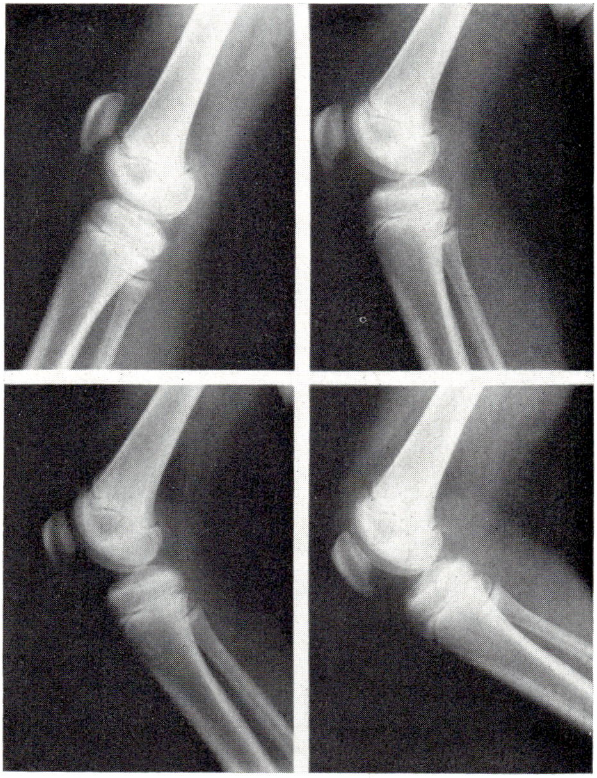

Fig. 5–19. Successive radiographs are taken in varying degrees of flexion to determine the instant centers.

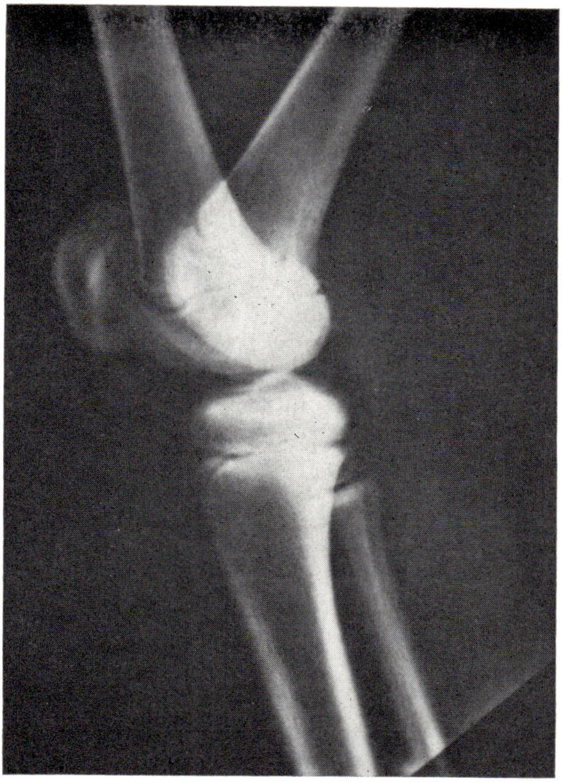

Fig. 5–20. Superimposition of successive radiographs. A larger displacement than usual is used for the purpose of illustration.

ments on the plane of the x-ray film. The perpendicular bisectors of the first (1,2) and last (4,5) set of vector displacements are shown in Figure 5–21B to intersect in centers C_{12} and C_{45}. By labeling the vector displacements X_{12}, X_{23}, etc., and Y_{12}, Y_{23}, etc., the intersection

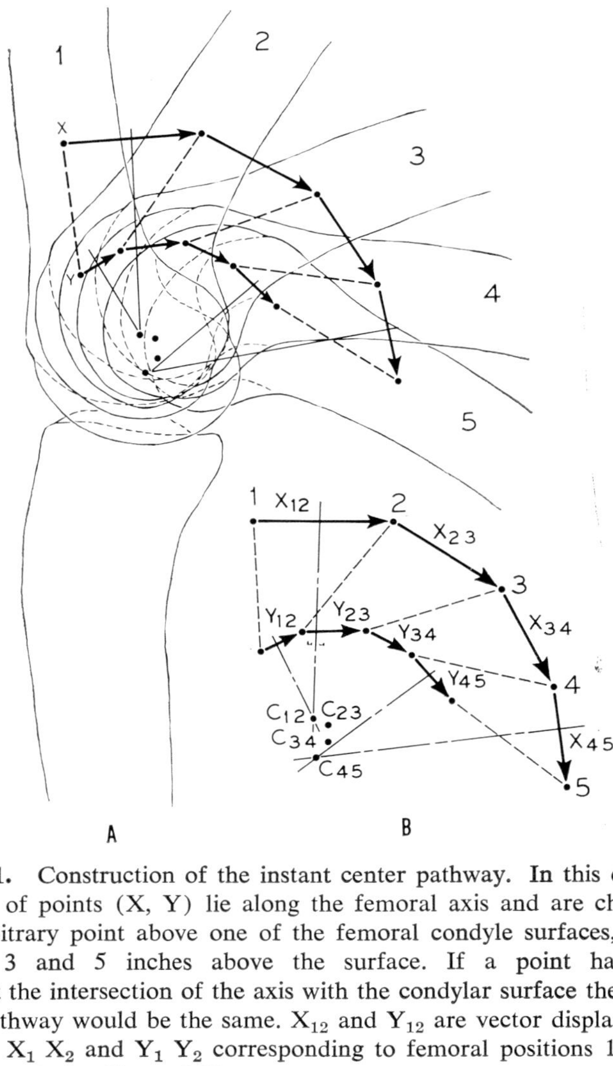

Fig. 5–21. Construction of the instant center pathway. In this example the pairs of points (X, Y) lie along the femoral axis and are chosen at some arbitrary point above one of the femoral condyle surfaces, in this instance 3 and 5 inches above the surface. If a point had been chosen at the intersection of the axis with the condylar surface the instant center pathway would be the same. X_{12} and Y_{12} are vector displacements of points X_1 X_2 and Y_1 Y_2 corresponding to femoral positions 1 and 2. C_{12} is the pole for X_{12} and Y_{12}.

of successive perpendicular bisectors may be noted. Thus C_{12} signifies the intersection of the perpendicular bisector of displacement 1 with the perpendicular bisector of displacement 2. Thus C_{12}, C_{23}, C_{34}, and so on form a continual set of points through which a curve is drawn. This curve is the location of the instant centers of rotation of the femur about the tibia and is one of the set of poloids. Figure 5–22 shows the relationship of the velocity instant centers to the displacement instant centers. The question that arises naturally from this discussion is: What is the instant center path of the tibia relative to the femur? It may be answered in one of two ways. The whole procedure outlined above may be repeated by fixing the femur and superimposing the pictures of the relative motion of the tibia in successive stages about the femur. There is, however, a second method of obtaining this curve, and this will now be illustrated. It will be remembered that the concept of the instant center implies that the instant center is the pivot point of relative motion between the two bodies. In effect this center of relative motion belongs to both bodies. In the above example, the instant center was only marked on the tibia. However, an alternate method is to mark the instant center on the femur as well. In this regard the instant center may be thought of as a pin which pierces both bones and thus acts as a hinge (Fig. 5–23). The concept of the extensibility of a body must be employed here since the piercing point of the imaginary pin on the tibia actually lies outside of the actual boundary of the tibia. As the two bones rotate to a new position, the instant center must change. Therefore, the imagi-

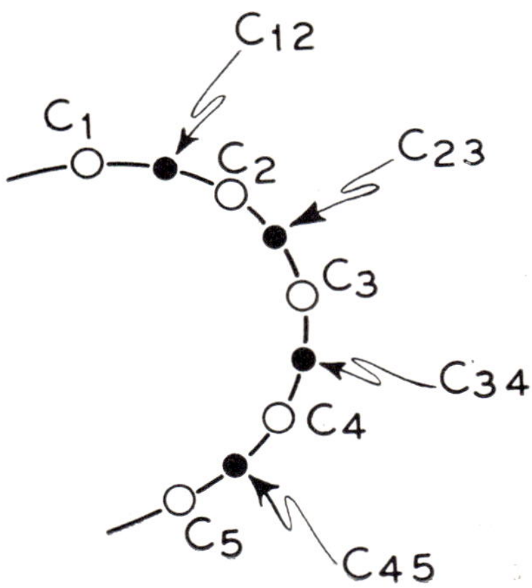

Fig. 5–22. The displacement centers lie between the instant centers for velocity.

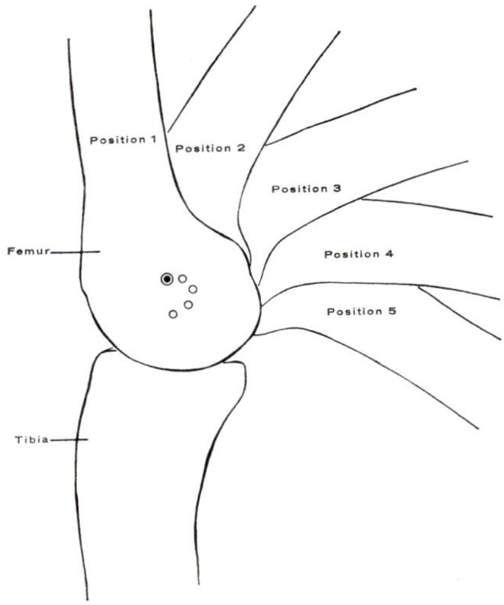

Fig. 5–23. Generation of paths (from position 1 through position 5) of the instant centers with respect to both the femur and tibia. These curves are called poloids.

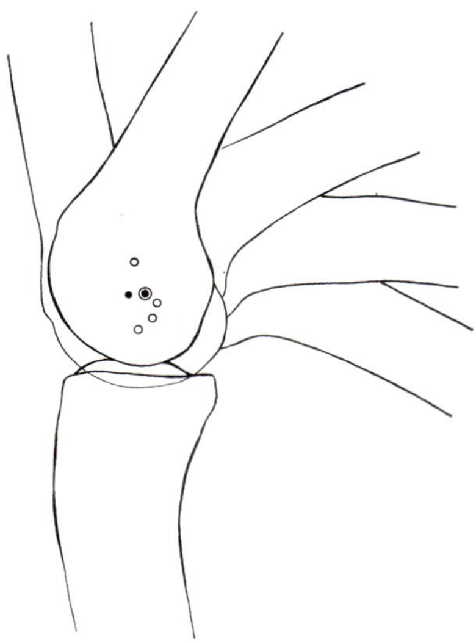

Fig. 5–24. The open circles are the instant centers for positions 1, 2, 3, and 4 with respect to the femur. The solid circle is the instant center for position 1 with respect to the tibia. The large open circle with the enclosed solid circle is the new instant center for position 2.

nary pin must be removed from its present location and inserted in a new location (Fig. 5–24). The imaginary hole that the pin leaves, however, can still be seen. Notice that when the second imaginary pin is inserted to act as the instant center between positions 2 and 3, the two holes made by the first pin have separated in space. In the previous example we left only the imaginary hole as it appeared in the tibia. In this example, both holes are left. The same procedure is repeated for each instant center necessary to produce the total range of desired motion. Note carefully that each instant center produces a hole or location on both the tibia and femur. In Figure 5–25 the successive holes or positions on both the tibia and femur are connected to form a continuous curve. This curve as shown on the tibia is identical with the one achieved in the previous example. The curve as formed on the femur is identical to one produced if the femur is held fixed, the tibia rotated about it, and the analysis performed as in the first example. These two curves form a complete picture of the relative motion which exists between the femur and tibia in normal knee motion as projected on the plane of the x-ray film.

These two curves are polodes which have an extremely interesting property. If each curve is reproduced as a three-dimensional solid and fixed to its respective bone, the motion that exists between the two curves during the normal knee function is that of pure rolling. That is, if the solid surfaces as shown in Figure 5–25 are fixed, one to the tibia and one to the femur, with the surfaces skewed to allow terminal tibial rotation, an analysis of motion as the knee goes through extension would show that the contact points between the two curves have zero relative velocity. This follows directly from the definition of instant centers. The contact point between the two curves is actu-

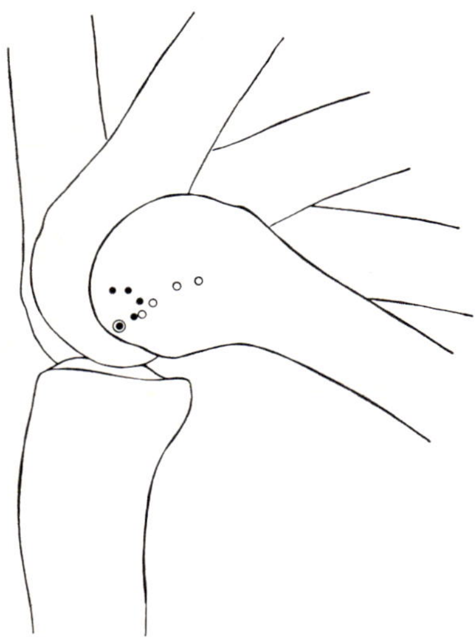

Fig. 5–25. The solid circles are the instant centers with respect to the tibia for positions 1, 2, 3, and 4. The open circles are the instant centers with respect to the femur for positions 1, 2, 3, and 4. The large open circle with solid circle enclosed is the instant center for position 5.

ally the instant center of motion between the two bodies, and, as such, there is zero relative velocity between the two curves at the point of contact. Thus, from the definition of rolling, it follows that the two curves are always in rolling contact as the knee extends and flexes. Because of this, these two surfaces undergo no sliding and, therefore, no frictional losses.

An examination of the paths of instant centers taken together with the successive relative positions between the tibia and femur can lead to a more complete understanding of the mechanics of joint motion. In Figure 5–26, a particular instant center together with the relative positions of the femur and tibia are shown. Superimposed on the figure are the directions of the velocities of three points on the femur. Remember that the tibia is held in a rigid position in this analysis. The directions of the velocities are obtained by drawing perpendiculars to the lines from the instant center to the three points on the contact surface of the joint. Each velocity is tangential to the contact surfaces. This indicates that the joint surfaces at these three particular points are neither separating nor tending to compress. The joint is sliding with a relatively free and normal motion. In Figure 5–27 a second joint is illustrated. Again, a particular position during the extension cycle is chosen and the instant center for this position is illustrated. Three joints on the joint surface are again chosen and the direction of the relative velocity of these points is obtained in the same manner as described above. In this case, however, it will be noted that the directions of the velocities do not lie tangentially to the surface of the joint. Instead, one of them seems to be tending to penetrate the joint surface while the other seems to be tending to

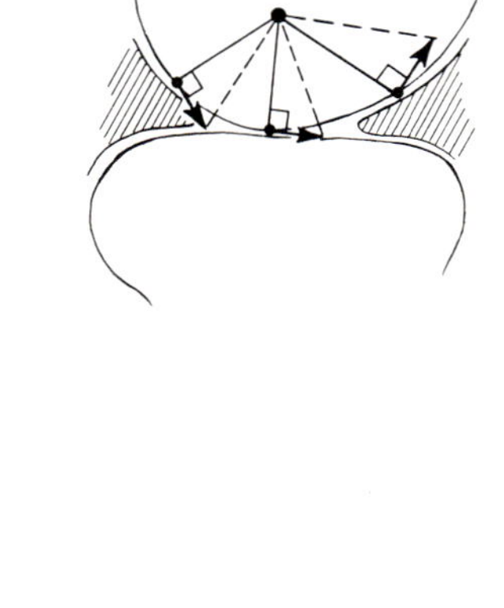

Fig. 5–26. If the instant center does not lie on the contact point, a relative velocity is seen to exist between the two contact surfaces. These surfaces are said to be in sliding contact.

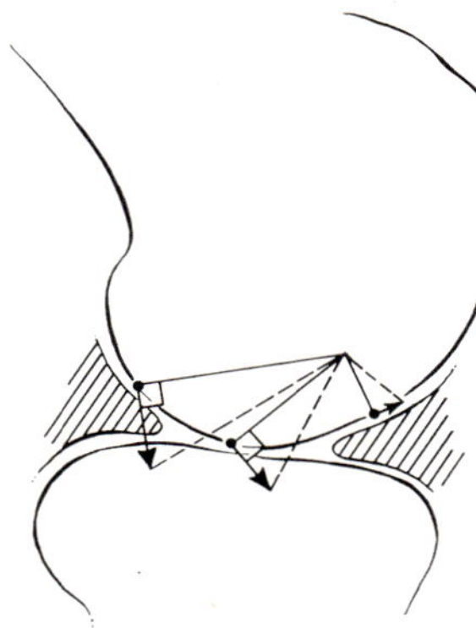

Fig. 5–27. The velocity of three surface points rotating about an abnormal instant center are shown. One point tends to penetrate the other joint surface.

separate the joint surfaces. The instant center has been displaced from its normal position because of a meniscal tear resulting in an abnormal surface reaction. The kinematic analysis indicates that excessive joint wear and trauma may occur because of the irregular kinematics induced into the knee by the internal derangement.

It has been well-documented that normal motion involves a moving instant center of rotation. Therefore, it is unreasonable to expect that normal motion may be forced or created by the use of a well-fitted, single-hinge knee brace. If there is sufficient clearance or play between the elements of the brace and the leg, normal motion may be permitted providing the joint itself is capable of producing it by being displaced. However, if it is desirable to force normal motion with well-fitted braces or hinged casts, a knee brace which produces a moving instant center of rotation is necessary. The science of kinematic synthesis allows the production of such types of mechanisms. There are several classes of mechanisms which can be constructed to produce the desired motion of the knee joint.

The first and probably the simplest of these is the mechanism produced by duplicating the shape of the polodes curves on a solid mounting and then mounting them on their respective bones, as described in Figure 5–25. A constraint which permits the two surfaces to roll over each other can then be fabricated. In this manner, desired motion of the two surfaces in the knee joint may be produced.

A second method consists of the design of a four-bar mechanism, which is a series of 4 links and 4 pivot points. This can be seen in Figure 5–28. Links 1 and 3 are fastened to the tibia and femur respectively. Links 2 and 4 produce the desired relative motion be-

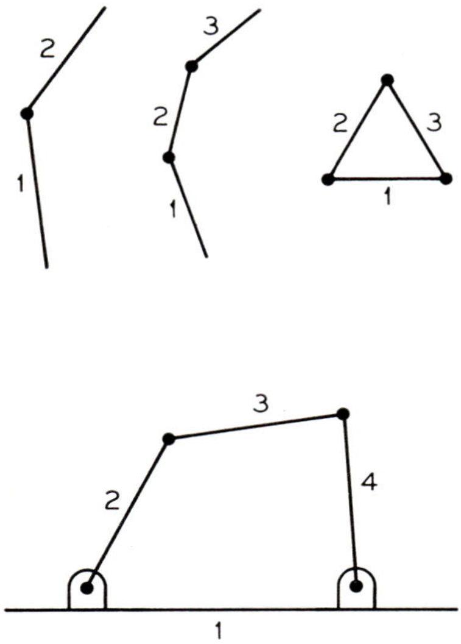

Fig. 5–28. Kinematic mechanisms. The four-bar mechanism is the simplest mechanism which has a determined instant center pathway.

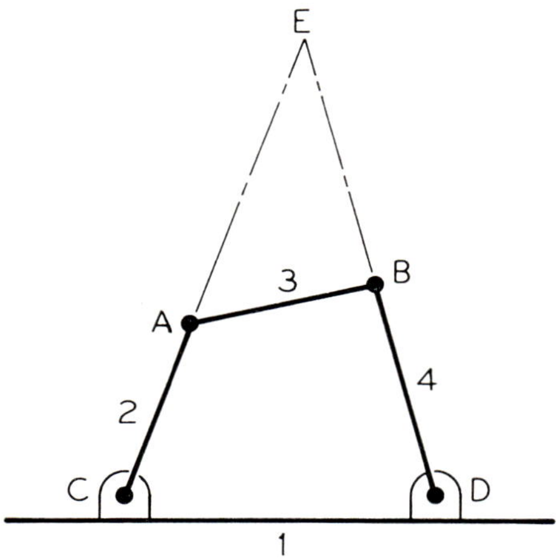

Fig. 5–29. The instant center for the position shown is at E.

tween links 1 and 3. Note that there are 4 pivot points instead of 1. If a constrained motion is to be produced, linkages with either 2 or 3 pivot points may not be used. In the case of a 2 pivot-point linkage, the motion is not constrained. Links 1 and 3 are free to move in almost any relative position. If three centers of rotation are used, then it is seen that a rigid triangular frame is created, and no relative motion may occur. Thus 4 pivot points comprise the simplest mechanism which may be produced beyond the concept of a single-hinge mechanism. Figure 5–29 shows a typical position of a four-bar mechanism. Link 1 is considered to be the fixed link; therefore the velocities of points A and B must be perpendicular to links 2 and 4 respectively. This is because the fixed pivot points of these 2 links are C and D. The velocities of A and B are therefore perpendicular to links 2 and 4. A and B are also points on link 3. Consider the velocity of point E on link 3. Point E is obtained by extending links 2 and 4 until they intersect. In a manner similar to that described earlier in this chapter, the orthogonal components of the velocity of E in the direction EC may be obtained. The procedure for obtaining this orthogonal component is to state that the component of the velocity of E along the direction EC is equal to the component of the velocity of A along the direction AC. Again, this is based on the principle of the inextensibility of a rigid body. Similarly, the component of the velocity of E in the direction ED can be determined. In determining components of the velocity of A in the direction AC, it is necessary to drop a perpendicular from the head of the velocity of point A to the line representing the direction AC. This line is actually coincident with the velocity of A and therefore A has no component of velocity in the direction AC. Similarly, B has no component of velocity in the direction BD. Therefore, E has no component of velocity in direction AC, nor does it have a component of the velocity

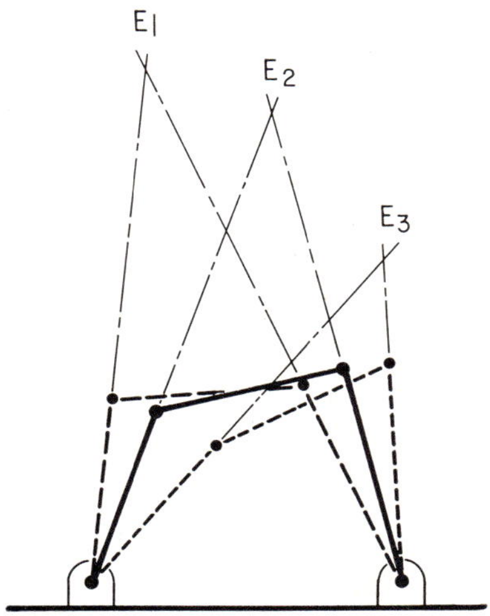

Fig. 5–30. Successive positions of the instant center of the top link occur at the intersection of the extension of the side links.

in the direction ED. The only velocity which can satisfy the condition of having the two zero components in these directions is zero velocity. Therefore, C is an instant center. It is the instant center of rotation of body 3 with respect to body 1. It is important to notice that this instant center is found merely by extending links 2 and 4. This procedure is repeated again in the next illustration (Fig. 5–30). Links 2 and 4 are extended, and a new instant center, called E_2 is shown. This process is repeated for a sequence of positions of link 3 and all of the instant centers thus obtained are connected to form a path of instant centers. In Figure 5–31 a four-bar mechanism has been drawn and its associated path of instant center of link 3 relative to link 1 is shown. Note that this path of instant center closely approximates the path of the instant center shown in Figure 5–22. Thus, this mechanism can be used to duplicate or force the knee motion produced by the patient whose x rays appear in Figure 5–19.

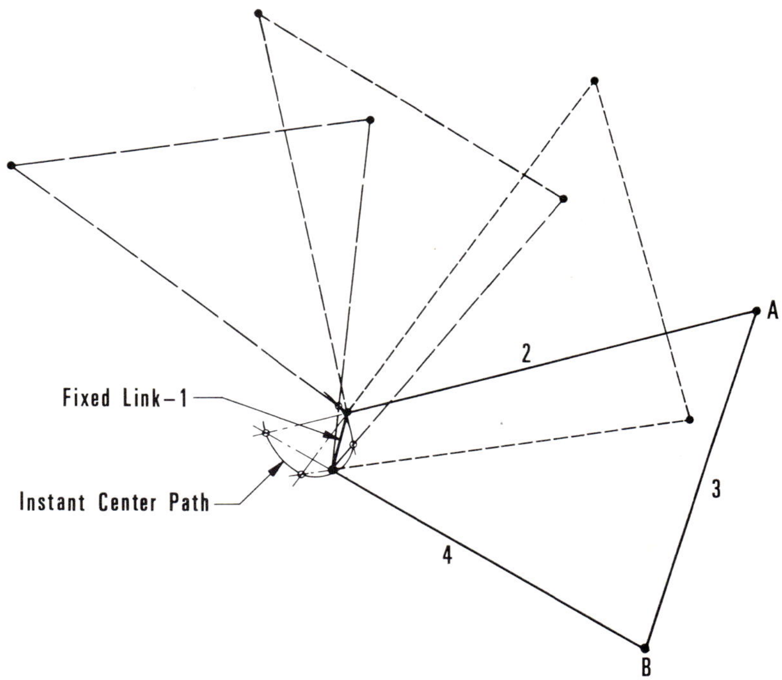

Fig. 5–31. A four-bar mechanism which approximates the path of the instant centers for the knee shown in Figure 5–21.

6. Lubrication

BEFORE studying lubrication in joints, it is necessary to review the concept of friction. Friction is a general term which describes the resistance two bodies develop when undergoing relative sliding motion. If body A in Figure 6–1 is trying to develop motion or is sustaining motion with respect to body B, friction develops. The friction force is measured in terms of pounds or kilograms.

The friction forces f_a and f_b in Figure 6–1 and the force F are equal in magnitude and the system is in equilibrium. This magnitude is a function of the relative velocity, as shown in Figure 6–2. At a very slow velocity, *e.g.,* 1 foot per year, the friction is high. It gradually falls to a minimum which is then relatively constant with respect to velocity. The contact surfaces, when examined under high magnification, are very irregular. The actual contact area may be a small

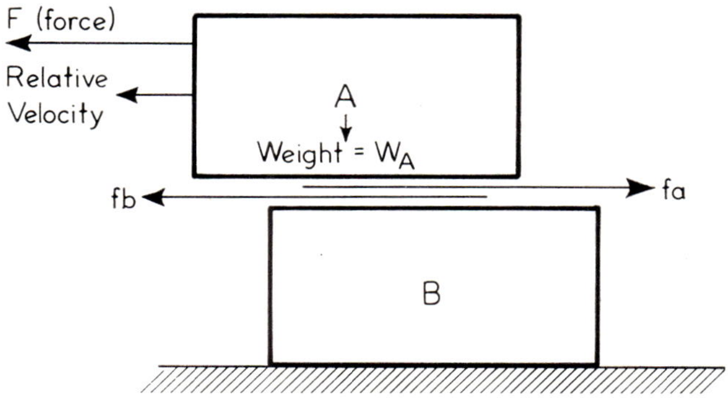

Fig. 6–1. Two sliding bodies. F is the force on body A, fa and fb are the friction forces developed at the interface of the respective bodies.

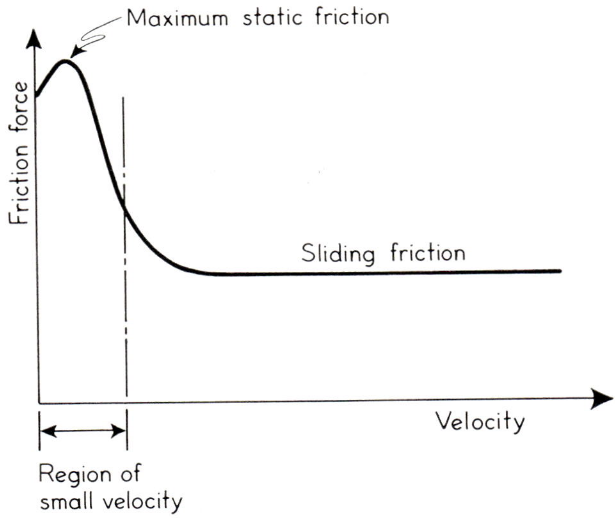

Fig. 6–2. Plot of friction force against the velocity. At very slow rates of velocity, the friction force is maximum with increasing velocity. The friction force decreases and remains relatively constant.

proportion of the total surface area. At very low velocities the contact surfaces of the two bodies develop points of attachment (cold welds) which must be disrupted.

Solid bodies have been found to develop friction force proportional to the normal force between the two bodies. In Figure 6–1, if the normal force between bodies A and B is equal to the weight of body A (designated W_A), the friction force is given by the formula

$$f_a = W_A \nu$$

in which ν is the proportionality constant, called the coefficient of friction. The value of this coefficient is seen to be dependent upon velocity only for very small velocities (Fig. 6–2).

The frictional effect between fluids is slightly different from that between solids. Two discrete fluid bodies do not rub against each other. Rather, there is relative motion between very small particles of fluid. Unlike the forces of solid friction, fluid friction forces are extremely sensitive to velocity (or more accurately, as shall be shown, sensitive to the velocity gradient). To illustrate this effect, recall the example in Chapter 2 which examined the relative resistance between fluid particles.

If the velocity of the fluid particles inside a moving laminar stream is observed, the results are as shown in Figure 6–3. The velocity of

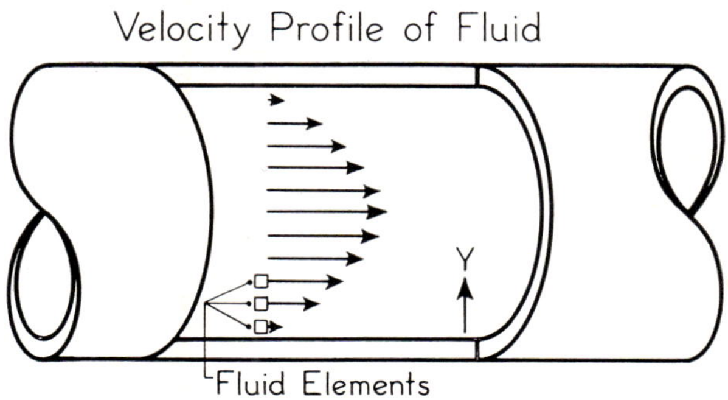

Fig. 6–3. Velocity profile of fluid in laminar flow. The fluid particles at the wall have zero velocity. Y is the distance of a particle from the wall.

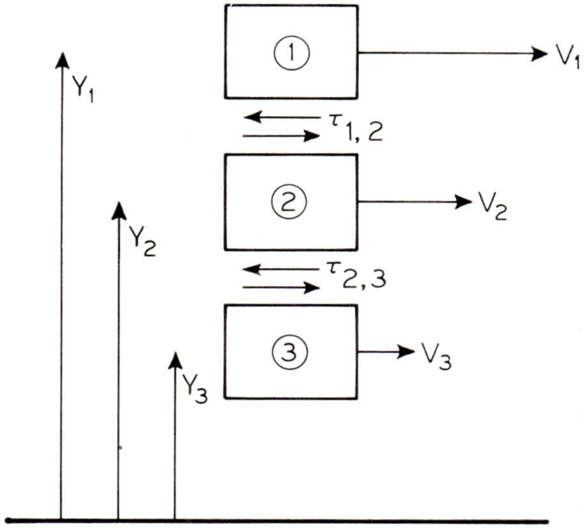

Fig. 6–4. Three adjacent particles taken from the flow stream in Figure 6–3. V_1, V_2, and V_3 are the velocities. Y_1, Y_2 and Y_3 are the positions of the particles in the stream. $\tau_{1,2}$ and $\tau_{2,3}$ are the shear stresses developed between adjacent particles.

the particles at the wall is zero. This value is attributed to friction and adhesion. The velocity of each particle is different. While those particles touching the wall have zero velocity, those in the center of the flow stream have maximum velocity. There is a gradation of velocity of fluid particles from the wall to the center of the stream. If three adjacent "particles" are observed (Fig. 6–4), they will be noted to have slightly different velocities. The difference in velocity between body 1 and body 2 is $V_1–V_2$. The distance between these two bodies is $Y_1—Y_2$. The velocity gradient is $\frac{V_1-V_2}{Y_1-Y_2}$.

The upper body (1) drags over the middle body (2), causing the shear force 1,2 as shown. However, the middle body is traveling faster than the lower body (3) and is therefore creating the shear force $\tau_{2,3}$. It is found that the frictional shear force between the upper and middle bodies is given by:

$$\tau_{1,2} = \frac{V_1 - V_2}{Y_1 - Y_2} \cdot \mu$$

That is, the shear is proportional to the change in fluid velocity per distance between particles, *i.e.*, the velocity gradient. The constant of proportionality, μ, is called the viscosity. The two bodies (1) and (2) are very small and, therefore, close together. The term $\frac{V_1-V_2}{Y_1-Y_2}$ is mathematically written as $\frac{dV}{dY}$ (derivative of velocity with respect to distance) and is called the velocity gradient. If a surface separated by a film of fluid from another surface is given a velocity, then a velocity gradient is produced in the fluid film. Three such cases are shown in Figure 6–5. In each case the velocity distribution is linear as shown. The velocity gradient in a in Figure 6–5 is 1 mm/sec/10 mm, which is equal to 0.1 sec^{-1}. The velocity gradient can be made large (as in c in Figure 6–5) by keeping the distance between two surfaces small. This produces a small denominator. It can also be made large by increasing the velocity between the two surfaces, as in b in Figure 6–5, by producing a large numerator.

The equation for shear stress in a fluid can be written as

$$\tau = \frac{V_1-V_2}{Y_1-Y_2}\mu \text{ or } \tau = \frac{dV}{dY}\mu.$$

The units of μ are $\frac{kg \cdot sec}{mm^2}$. This is a rather awkward combination of

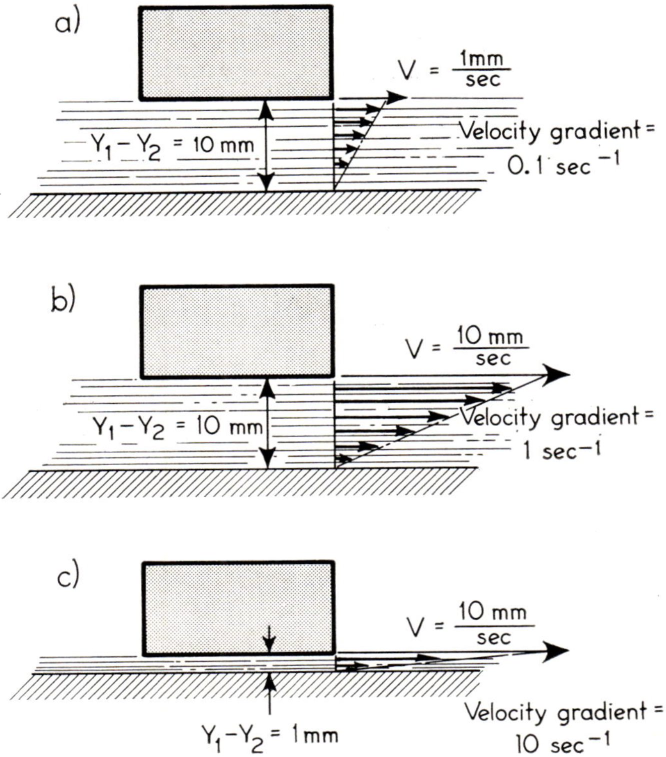

Fig. 6–5. Three velocity gradients (a) (b) (c) produced by varying either the film thickness or velocity.

units, so, by convention, this unit is called a poise. For standardization, one centipoise is taken to be the viscosity of water at 20° C and atmospheric pressure.

Viscosity, like the coefficient of friction, is not a direct measure of the actual friction force. Rather, they are both material properties which must be considered in combination with another variable of the system before frictional forces can be calculated. In a system of solids, the second variable is normal force while in a liquid system, the second variable is velocity gradient. Viscosity will also be shown to play an important role in determining the suitability of a fluid for use as a lubricant. Just as the moduli of elasticity of various materials differ, so do the viscosities of fluids differ. Thus, the viscosity of a lubricating oil used in an internal combustion engine may be 5000 times that of water. It is also of interest to note that while some fluids (*e.g.*, water) have viscosities which remain rather constant even under extreme pressure, temperature, and velocity gradient changes, other fluids such as oil experience large changes of viscosity with changes of temperature. Synovial fluid, however, undergoes large changes in viscosity, not only with changes in temperature, but also with changes in velocity gradient. While the viscosity of water remains at 1 centipoise (at ambient temperature and pressure) with changing velocity gradient, the viscosity of synovial fluid varies from 100 poise at a velocity gradient of 0.1 sec^{-1} to 1 poise at a velocity gradient of 100 sec^{-1} in Figure 6–6. A fluid whose viscosity is independent of the velocity gradient (*e.g.*, water) is called a Newtonian fluid. A thixotropic fluid is one in which the viscosity varies with the velocity gradient.

To describe the complex and varied phenomena of lubrication the following model will be observed (Fig. 6–7). The velocity of the weight relative to the ground will be allowed to vary from zero to

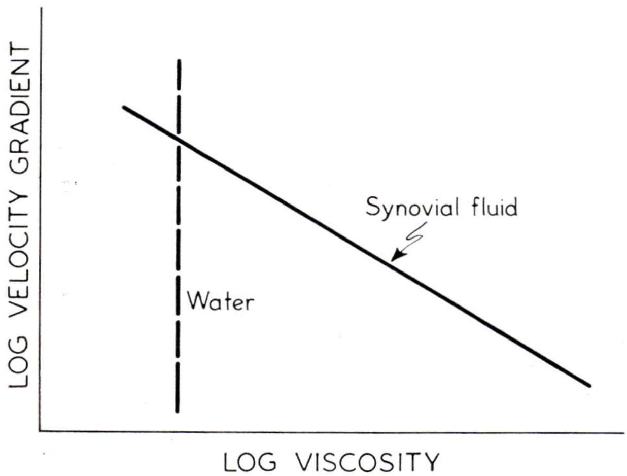

Fig. 6–6. The relationship between the velocity gradient and the viscosity for two fluids. A change in the velocity gradient of water produces no significant change in the viscosity. Changes in the velocity gradient for synovial fluid produce significant changes in the viscosity. On the log-log plot, these changes appear to follow a straight-line relationship.

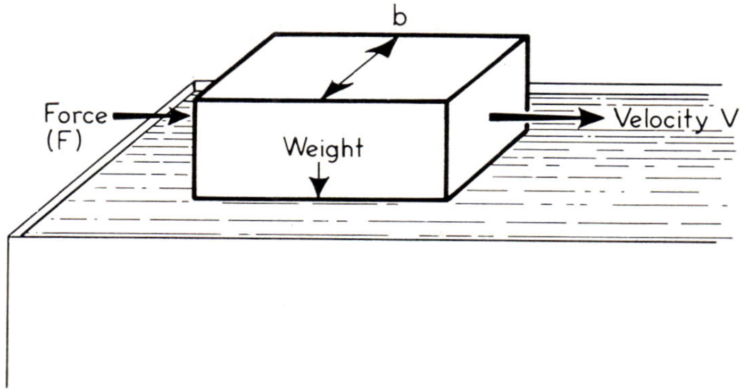

Fig. 6–7. Model lubrication experiment. The weight is being pulled across a surface covered with a lubricating fluid. A force F is being applied to the weight, which has a width of b. The resulting velocity V and the geometry of the lower surface of the weight in relation to the table can be observed.

some large value. It will be assumed that there is always an abundant supply of Newtonian fluid available at the contact surfaces. Observation of the model will show the force necessary to sustain the velocity as well as the micro- and macrogeometry of the touching surfaces. The force (F) versus velocity (V) data will be plotted as the parameters $\frac{F}{(W)}$ and $\frac{\mu Vb}{(W)}$. Combinations of parameters such as these are termed nondimensional parameters since they have no units or dimensions; for example:

$$\mu V \cdot b \cdot \frac{1}{W} = \frac{kg \cdot sec}{mm^2} \cdot \frac{mm}{sec} \cdot \frac{mm\,1}{kg} = \text{NO UNITS}$$

Nondimensional parameters are considered as single experimental variables so that the results may be presented as concisely as possible. Each element in the nondimensional parameter affects the experiment in an identical manner, $\frac{1}{W}$ being considered as an element rather than W. Thus, in an experiment in which V, b, and W are held constant, the effect of a variation of μ or F is as shown in Figure 6–8. Another experiment holding all variables except μ constant

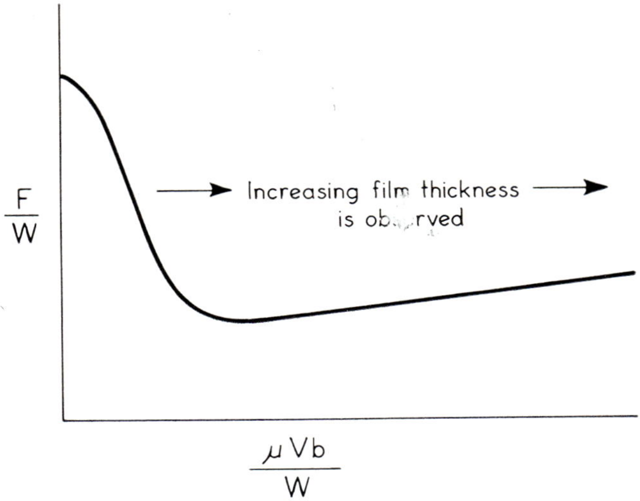

Fig. 6–8. Plot of nondimensional parameters observed in the experiment illustrated in Figure 6–7. F is the force necessary to move the weight at velocity V. W is the weight of the block, b is the width of the block, and μ is the viscosity of the fluid in the interface.

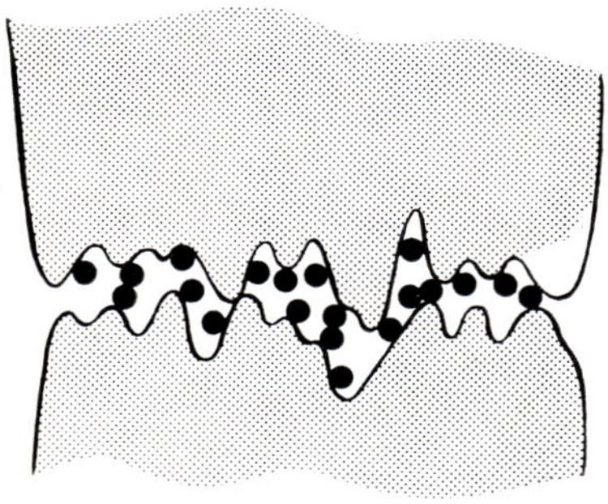

Fig. 6–9. Boundary lubrication. Molecules of lubricating fluid adhere to and fill in irregularities of the contact surfaces.

produces the same results, but the abscissa is relabeled μ. And in the same manner, variation in b also produces identical results. In general then, Figure 6–8 represents the total results of all of the above experiments.

The results of this experiment can be interpreted partially in terms of the film thickness produced on the model. At the extreme left of the graph in Figure 6–8 the film thickness is of the magnitude of several molecules.

Lubrication results from the "slipperiness" of the molecules and from the filling in of the irregularities of the rubbing surfaces (Fig. 6–9). Graphite and other slippery solids as well as oil, cream, ink, and other fluids are effective lubricants in this region. There are important defects involved with this mode of lubrication. Wear of the contact surfaces occurs when the lubricating process is unstable. If the surface is heated as the result of the work produced by friction force F, the viscosity μ of most fluids tends to be lower. Figure 6–8 shows that if the variable $\frac{\mu Vb}{W}$ is lowered by lowering μ, a still larger force is required to overcome friction. This in turn will produce more heat due to the increasing frictional work. Thus, the system goes from bad to worse. This lubrication is termed boundary-layer or thin-film lubrication.

As the velocity of the block is increased, the experimenter will note that the block tilts and a wedge-shaped layer of lubrication fluid is formed (Fig. 6–10). There is now no contact between the block and the underlying support. This is termed hydrodynamic or thick-film lubrication. An entirely different lubrication mechanism is now present. A hydrodynamic force is developed and the fluid thickness is observed to be at least several 100 times the molecular size of the fluid. In addition, the film thickness increases as the operating point is moved to the right on the abscissa.

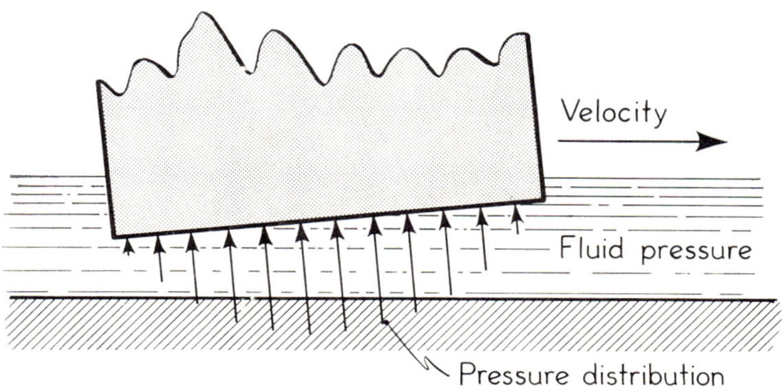

Fig. 6–10. An enlarged examination of the geometry of the lower surface of the weight shown in Figure 6–7. At sufficiently high velocities, the weight tips, forming the wedge shape as shown. Because of the viscous properties of the fluid a pressure will be created within the fluid to support the weight.

Since a force is being exerted on the moving block to sustain its velocity, energy is being put into the block and, hence, into the fluid. The energy is dissipated within the fluid as heat, but in the process a pressure is created in the fluid layer. This pressure is due to two factors: the first is the viscous property of the fluid; the second is the wedge shape between the moving surfaces. The resulting distribution of pressure is shown in Figure 6–10. It is important to note that the mass is supported by a hydraulic pressure. It never touches the lower surface and no wear occurs.

A small relative velocity must exist between the surfaces before hydrodynamic lubrication can take place. As the velocity increases, the friction force also increases owing to the increased viscous shear drag.

Consider the effect of a thixotropic fluid on the same system. The experimental data are illustrated in Figure 6–11, in which μ_0 is the viscosity of the fluid at some particular velocity gradient $\left(\frac{dV}{dY}\right)$. It is not logical to include the actual viscosity in the nondimensional abscissa function. An experiment can no longer be performed with only a change in viscosity because the film thickness of the fluid increases as the abscissa parameter increases; hence, the velocity gradient changes. Since the fluid viscosity is dependent upon velocity gradient, and thus on film thickness, it must change. The fluid viscosity is therefore chosen at an arbitrary velocity gradient to be used in the nondimensional parameter.

The experimenter will observe two differences between the results of this experiment and the previous experiment with a Newtonian fluid. The friction force does not increase as rapidly with increasing velocity and the film thickness is maintained at a more constant value. The reason for these two differences is evident. The actual viscosity of the fluid decreases with increasing velocity gradient, and thus the expected change due to increased velocity is not fully realized. The change in slope of the thick-film portion of the curve for the thixotropic fluid in Figure 6–11 is the result. Since the thixotropic fluid has a much higher viscosity at lower velocity gradients, it is able to enter the region of hydrodynamic lubrication at much lower velocities. Thus, the left side of the curve is shifted further to the left. The composite result is that a thixotropic fluid allows hydrodynamic lubrication at lower relative velocities and allows lower coefficients of friction at higher relative velocities.

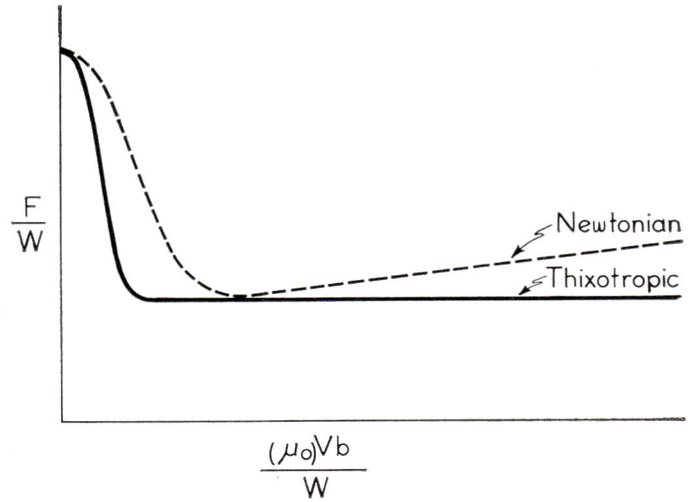

Fig. 6–11. The results of an experiment utilizing the conditions of Figure 6–7, but with a thixotropic fluid, are superimposed on the results of the same experiment with a Newtonian fluid. μ_0 is the viscosity of the thixotropic fluid at an arbitrary velocity gradient. Other parameters are as in Figure 6–10. The film thickness does not increase as rapidly with increasing velocities as when thixotropic fluids are used as the lubricant.

The application of these principles to the study of synovial joints is complicated by the material properties of the articulating surfaces. Cartilage is a viscoelastic, porous material. Therefore, the pressure distribution which exists on the articulating surfaces results not only from the static or possible hydrodynamic behavior of the fluid, but also from the viscoelastic interaction of the surfaces. It has been called an elastohydrodynamic system. The investigation of such systems is in its initial phases.

7. Design

THE process of engineering design requires a knowledge of several areas in material science, mechanics, and manufacturing technology. In addition, an orderly method of transforming the design objective into a physical model is required. The data that must be acquired in designing an appliance and the method of design and manufacture may be classified into seven categories. They are: (1) properties of appliance materials; (2) properties of biological materials; (3) kinematics; (4) kinetics; (5) lubrication; (6) design elements; and (7) cost analysis and manufacturing processes. This chapter concerns itself with design elements.

In designing devices, the basic consideration for the designer is the nature of the load to be borne by the device. The types of loading may be divided into four classifications. These are (1) tension; (2) compression; (3) torsion; and (4) bending. In the two simple categories of tension and compression, the strength of the device is directly proportional to its cross-sectional area. A modification of this statement is necessary in the case of compressive loading on long thin members (see Chapter 3). In this case, the possibility of buckling exists. The conditions leading to buckling are seldom encountered, however, in normal application of orthopaedic devices. The most common loading configurations of torsion and bending are often encountered. Appliances such as nails, plates, and intramedullary rods are subjected to loading configurations which produce bending moments at various sections of the appliances. Since the function of the device is twofold: (1) to provide rigidity of fixation, and (2) to support the applied load, there are several factors which must be examined in order to determine optimum section design.

In a device subjected to bending moments, the deflections produced are proportional to the moments applied and to the cube of the length of the device. The deflections are also inversely proportional to the area moment of inertia of the cross section. Anatomical considerations dictate the necessary length of the implant device. The loading configurations are the results of gravitational loads, muscle pulls, and ligamentous viscoelastic forces. The designer has one major design variable which he can control. This is the area moment of inertia and/or the polar moment of inertia of the appliance.

Generally, the loading configuration to which an implant device is subjected is not limited to one particular plane. There may be simultaneous forces applied in several planes. If this is the case, cross sections which are asymmetrical may not be as satisfactory as those which are symmetrical. For example, while an "I" cross section is extremely rigid in the direction shown in Figure 7–1, its rigidity or area moment of inertia is decreased by 95% if the load application is in the plane shown in Figure 7–2. By comparison, the round section shown in Figure 7–3, which has the same diameter as the height of the "I" section and the same cross-sectional area, has equal properties in all directions and is at the same time 85% as rigid as the "I" section.

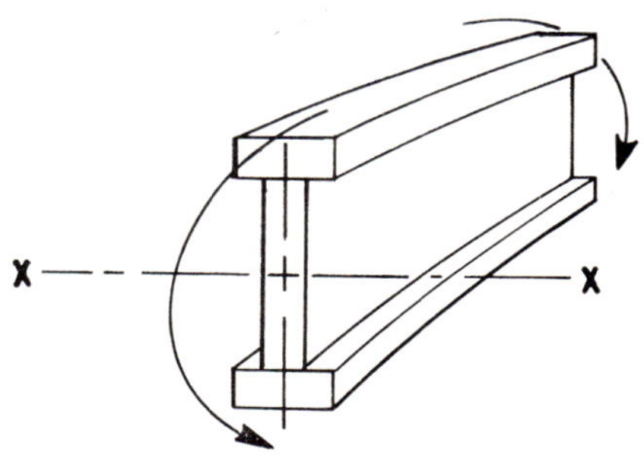

Fig. 7–1. The maximum area moment of inertia for the I-shaped cross section shown occurs when the bending axis is taken as the x-x axis. For this axis $I_x = 1.65$ in^4.

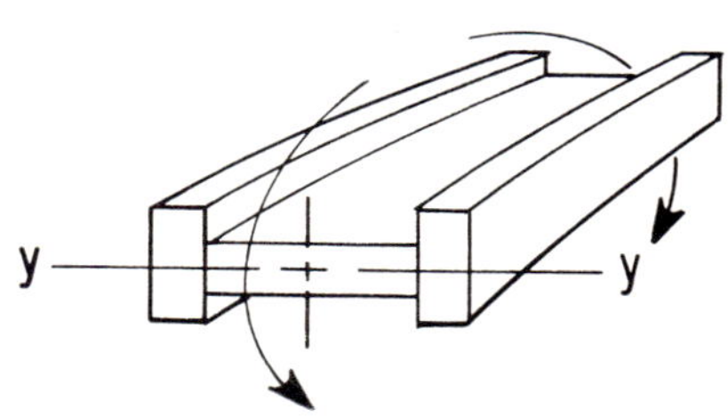

Fig. 7–2. When the same cross section as was shown in Figure 7–1 is oriented so that the bending axis is now the y-y axis, the resulting area moment of inertia is decreased by 95% to $I_y = 0.05$ in^4.

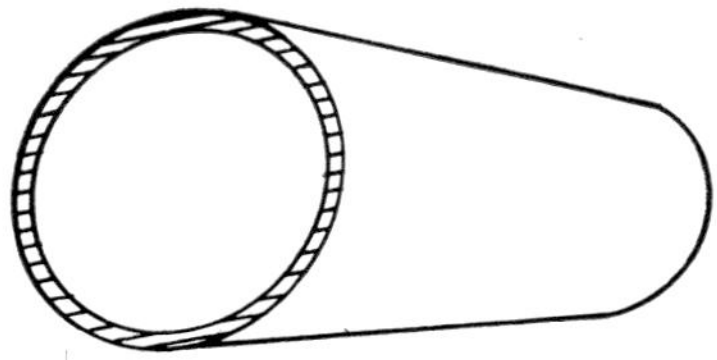

Fig. 7–3. The hollow circular section has the same cross-sectional area as the I section shown in the previous figures. Its area moment of inertia in all directions is equal to 1.4 in^4.

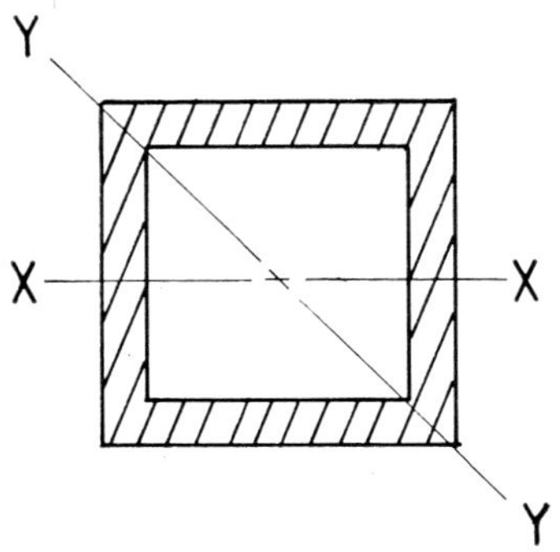

Fig. 7–4A. The hollow square section has area moment of inertia values in the x-x and y-y axis, which do not differ. If l is the length of the side, $I_{xx} = I_{yy} = 0.057\ I^4$.

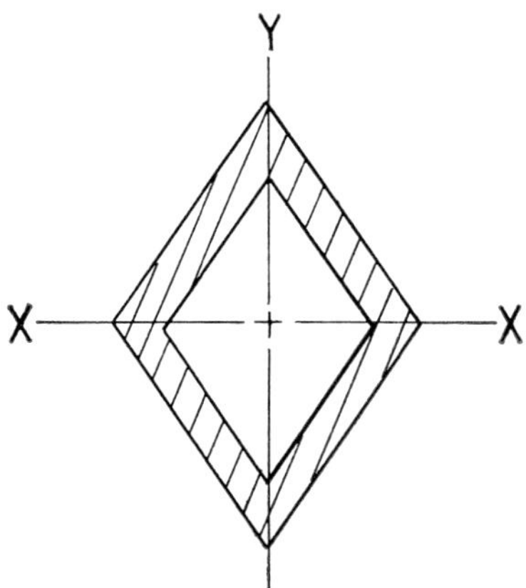

Fig. 7–4B. A cross section in the shape of a hollow parallelogram has values for area moment of inertia which differ only slightly. If l is the length of the side, $I_{xx} = 0.070\ I^4$ and $I_{yy} = 0.041\ 1^4$.

Sections which are almost symmetrical in all directions such as the square shape shown in Figure 7–4A or the diamond shape shown in Figure 7–4B have values of area moment of inertia which do not differ significantly in varying directions. This can be seen in the caption in each illustration.

In considering the design of appliances subjected to bending load, careful attention must be paid to the continuity of the shape of the cross-sectional area. Recall that, as presented in Chapter 2, the maximum stress produced in a section in bending was described by the formula

$$\sigma = \frac{M \cdot Y_{max}}{I_a}$$

In a device subject to bending loads, a sudden alteration of the section modulus (I_a/Y_{max}) anywhere along the length of the device will result in a section of greatly reduced strength. For example, as shown in Figure 7–5, if the nail plate is constructed with a nail section of large section modulus but a nail-plate junction of low section modulus, large stresses will be developed in the nail-plate junction. This will result in a bending failure, as shown in Figure 7–6.

Design considerations for members which must resist torsional loading are similar to those required for sections resisting bending loading. In the case of torsional loading, the angular deflection experienced by the member is given by the formula

$$\theta = \frac{TL}{JG}$$

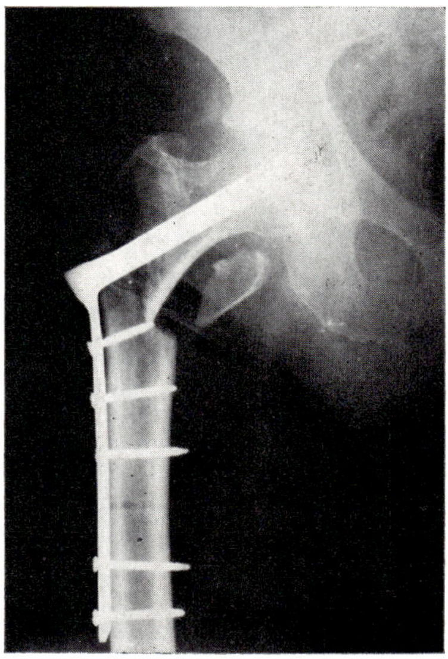

Fig. 7–5. A nail plate which is designed with differing section moduli for the plate section and for the nail section.

in which θ is the angular deflection, L is the length of element, J is the area polar moment of inertia of the cross section, and G is the modulus of rigidity. Again, the only variable over which significant control may be exerted is the area polar moment of inertia of the cross section.

As previously stated in Chapter 2, there are two general classifications of cross sections which need be considered in a discussion of torsional loading. They are the open section and the closed section.

In the closed section, a symmetrical section which places the material as far from the center as possible is the most efficient design.

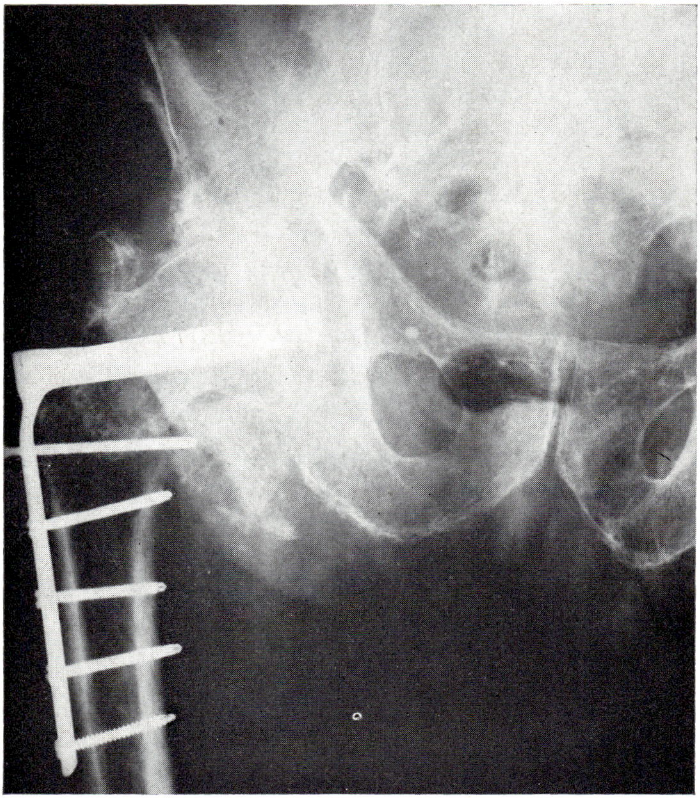

Fig. 7–6. The same type of nail as shown in Figure 7–5 failed by bending at the junction of the high section modulus portion and the low section modulus portion.

Open sections possess only a fraction of the rigidity of closed sections of similar geometry. For example, rigidity of the clover-leaf open section as shown in Figure 7–7 is 14% of the rigidity of the diamond shape nail shown next to it. Both nails have the same diameter.

Consideration of the maximum strength of devices suggests that the proper value for the polar moment of inertia or the area moment of inertia is the maximum obtainable value consistent with the biologically imposed restrictions on size. This is actually the case for only a limited number of applications. The optimum value for the area moment of inertia need be a maximum only when the implant device is not used as an assistive device, that is, when it must bear the total load of the biological structure. Examples of devices which bear the total load include femoral head prostheses, and artificial knee and elbow joints. Nail plate and intramedullary rods require a different criterion for determining the most suitable value of the area moment of inertia or the polar moment of inertia because they usually do not bear the whole load. To determine the optimum value of these variables, it is necessary to investigate closely the desired function of these types of elements throughout their useful life range. The initial and basic purpose of these devices is to provide a means of load transmission across fractured bony members. In performing this function, it is usually desirable to maintain some minimal amount of relative motion between the contiguous surfaces of the fracture site. Thus, it might appear desirable that the stiffness of the internal fixation device be at a maximum. Consider, however, the function of the implant device after the fracture has healed. If the device is to be removed at this time, it may indeed be made so as to provide maximum stiffness or rigidity. If the device is to remain implanted after healing has occurred, then, ideally, the device should be designed so that it no longer bears any load. This is desirable both from the point of view of having the biological structure bear its normal load

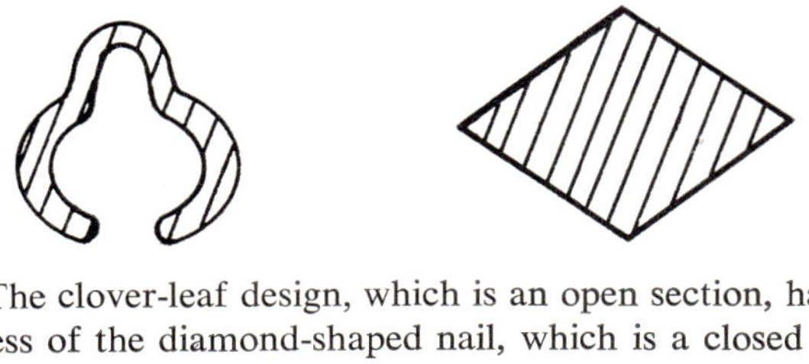

Fig. 7–7. The clover-leaf design, which is an open section, has only 14% of the stiffness of the diamond-shaped nail, which is a closed section.

and for the purpose of preventing a fatigue failure in the implant device. During the patient's remaining life span, the implant device will be subjected to cyclic loading tens or hundreds of millions of times. If the implant device is carrying a significant load, it may fail from fatigue. The amount of load carried by the implant device after the fracture has healed can be determined by comparing the relative rigidity of the implant device to the bone to which it is attached. This is illustrated in the following example.

A steel rod and a rubber rod are attached between two plates and a weight is placed on one of the plates as shown in Figure 7–8. The steel rod elongates very slightly due to the stress that it develops in resisting a portion of the applied weight. The steel rod stretches about $1/10{,}000$ of an inch and the rubber rod, also attached between the plates, stretches the same amount. How much of the load is being supported by the steel and how much is being supported by the rubber? When the steel rod elongates $1/10{,}000$, it is resisting with a force of 495 pounds. This may be calculated by first determining the strain in the rod.

$$\text{Strain } (\varepsilon_x) = \frac{\text{Change in Length (.0001) in}}{\text{Original Length (2 in)}}$$

$$\varepsilon_x = .00005$$

The stress in the steel rod may be calculated from

$$\text{Stress } (\sigma_x) = \text{Modulus of Elasticity (E)} \cdot \text{Strain} \cdot (\varepsilon_x)$$

$$\sigma_x = 30{,}000{,}000 \text{ psi} \times .00005$$

$$\sigma_x = 1500 \text{ psi}$$

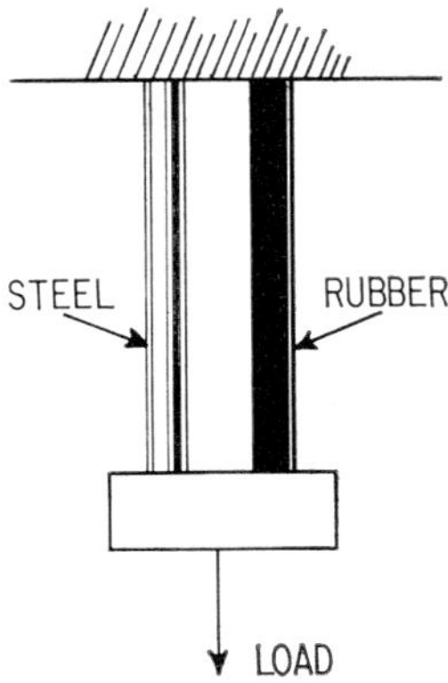

Fig. 7–8. An idealized experiment is performed using a rubber rod and a steel rod to support a given load.

The load on the rod may be calculated from its area and the value of the above stress. When the rubber is elongated $\frac{1}{10,000}$, it resists with a force of approximately 5 pounds. Thus, the rubber rod would take up approximately 5 pounds. The steel rod would take up approximately 99% of the load and the rubber rod would take up only 1% of the load. If two elastic systems support the same load in a manner that causes both systems to deform equally, the load will be divided according to the rigidities of the systems. The more rigid system will support the greater load, the less rigid system will support the lesser load. Therefore, the designer should make the device sufficiently rigid so as to provide no more than the maximal tolerable amount of relative motion during the healing process but not so rigid as to force the implant device continuously to carry the load after healing has taken place.

Because of the change in function that an implant device undergoes after the fracture site has healed, another design possibility is apropos. This alternative is to construct an implant device of varying rigidity; that is to say, the section or geometry of the implant device is not constant over the length of the implant device. An example of this would be a tapered plate, as shown in Figure 7–9. In this design, the section between screw holes 3 and 4, where the fracture line lies, is constructed so as to provide maximal rigidity and minimal movement at the fracture site. The portions distal to this section taper very quickly and therefore are significantly less rigid than the central section. A design of this type absorbs less load from the sections of the bone under the ends of the plate after the fracture has healed and therefore is less subject to fatigue failure.

Another advantage of the tapered design can be illustrated with the tapered nail-plate design shown in Figure 7–10A. Because of the tapered design, all of the sections such as 1, 2, and 3, have approximately the same ultimate load-carrying capacity; that is to say, they

Fig. 7–9. The design of an implant device may be a compromise between the rigidity necessary to provide adequate fixation during fracture healing and the flexibility necessary to allow the normal bone structure to bear the greater percentage of loading. The tapered plate provides rigidity in the area of the fracture. The rigidity of the plate ends is significantly lower than that of the center.

can resist the same load placed at the tip of the nail before they plastically deform. The reason for this is that the moment varies along the nail plate, being greatest at the nail-plate junction and least at the tip. The tapered design thus provides a greater strength of section

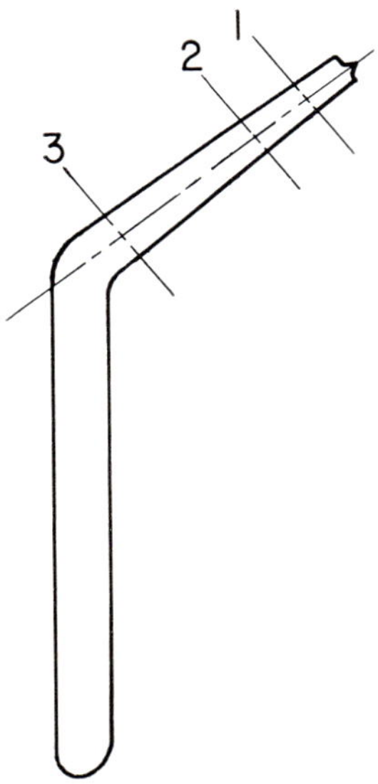

Fig. 7–10A. The tapered nail plate design allows a nail of constant strength. Each of the sections, such as 1, 2 and 3, produces the same maximum stresses when the nail is loaded at the tip.

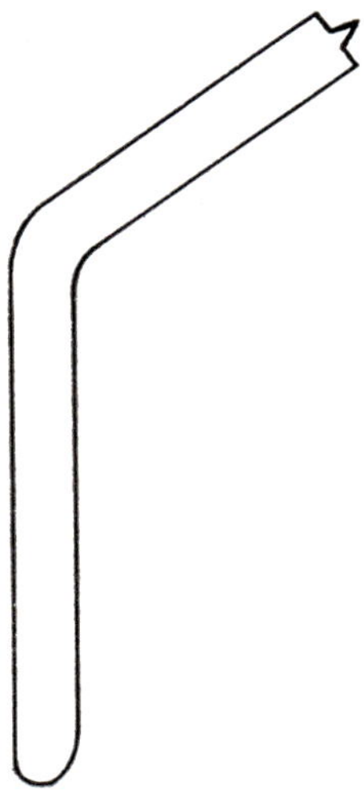

Fig. 7–10B. The strength of this nail is equal to that of the nail shown in **A.** Strength is used here to denote the maximum load-carrying capacity.

at the junction where it is needed, and a progressively smaller section moving proximally toward the tip of the nail which coincides with a progressively smaller bending moment. The load deformation curve of the tapered nail plate is shown in Figure 7–11. Superimposed on this is the load-deformation curve of nail B in Figure 7–10. The difference between these two nails is only that nail B does not have a tapered section. Both nails will withstand exactly the same load before permanent deformation occurs. Nail A has slightly less overall rigidity; however, nail A will absorb a larger amount of energy before failure occurs.

In considering failure of implant devices, two general types of conditions are seen. The first, as shown in Figure 7–6, consists of excessive permanent deformation of the implant. The second, as shown in Figure 7–12, consists of fracture through a section of the implant device. Both of these conditions occur because of the development of stresses within the implant device which are greater than the yield stress of the material. If the device is subjected to loads in excess of its design limits, the type of failure characterized by excessive plastic deformation will result. The second type of failure, however, can occur at loads below those normally considered safe for the appliance. This can occur because of the effect of stress concentrations or fatigue. Stress concentrations may appear in the design of an appliance in the form of small holes or sharp corners, as shown in Figure 7–13. Nicks or grooves may be produced during the insertion procedures if the appliance is improperly handled. In either case, the effect of a stress raiser is to increase the stresses in the material immediately adjacent

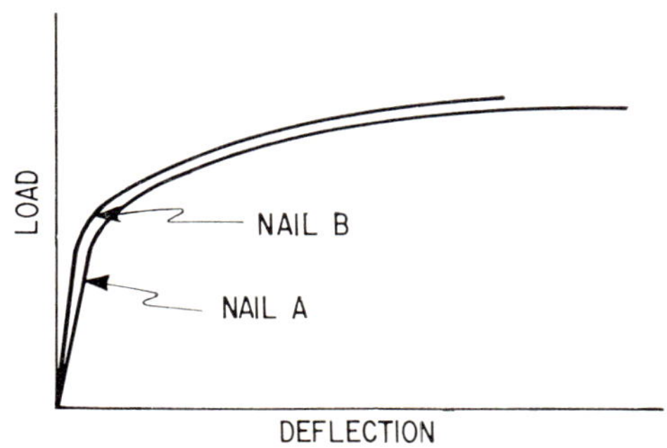

Fig. 7–11. Load-deformation curves for the nails shown in Figures 7–10A and B.

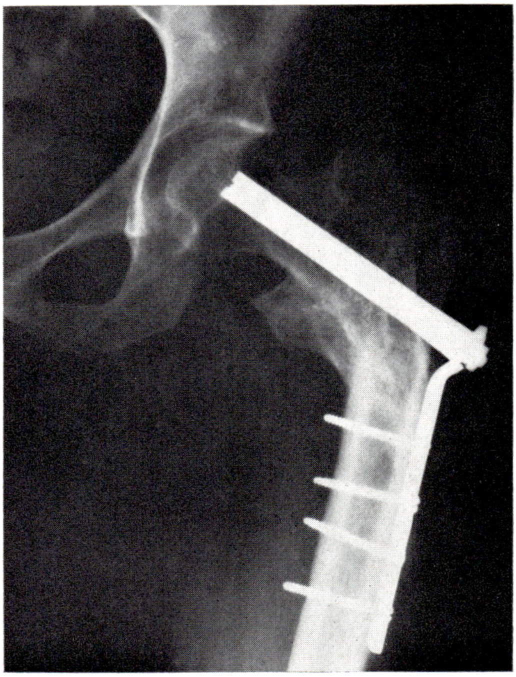

Fig. 7–12. Failure in an implant device due to fracture of the metal by the process of fatigue.

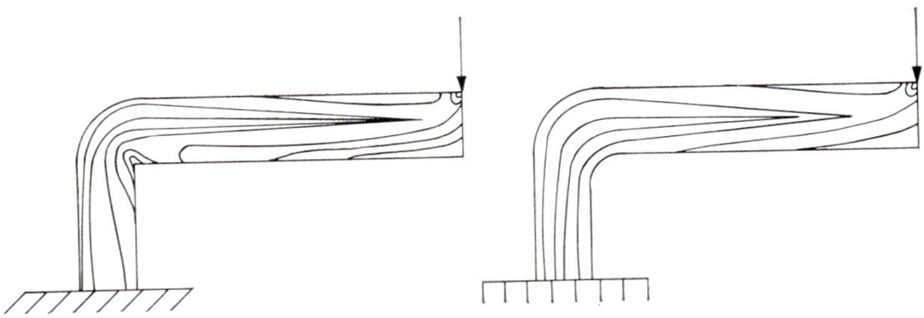

Fig. 7–13. When the beams shown above are subjected to a bending moment, significantly higher stresses are developed in the left-hand beam. The area moment of inertia of both beams throughout their lengths is equal. However, the left-hand beam possesses a sharp discontinuity or a corner which causes a local concentration of stress.

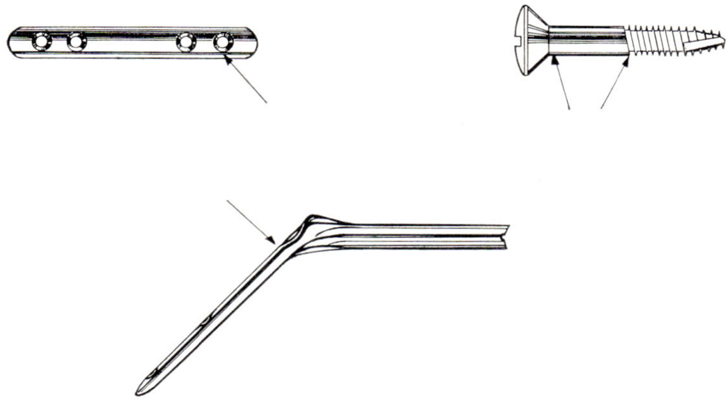

Fig. 7–14. Some commonly encountered stress rises.

to the defect. Any geometric configuration which produces a discontinuity in the geometry of the device is a stress raiser (Fig. 7–14). A stress concentration may take the form of suddenly changing sections, junctions between threaded and unthreaded portions, holes, scratches, inclusions within the material itself, or re-entrant corners such as the base of threads or the corner of a clover-leaf nail.

A simple analogy which is helpful to the designer in determining whether he has included stress concentrations in a particular appliance is to think of the stresses that exist in a member as analogous to the velocity of water flowing in a stream. In the example shown in Figure 7–15, a plate is put under a bending loading. The analogy to this configuration is a flow path shaped in the same geometry of the plate with water entering the path at some constant flow velocity. In the areas immediately adjacent to the obstruction to the fluid flow, large flow velocities are encountered. This is, of course, necessary since the areas are restricted in this region and, in order for the fluid to move through a restricted area, a higher velocity of flow is necessary. The velocity of flow is not, however, uniform across the restricted section. Rather, it is much larger in the areas immediately adjacent to the obstruction. In Figure 2–7, the stress concentration effect of a hole is visually demonstrated by a technique known as photoelasticity. The density of the black lines is an indication of the magnitude of the stresses. Note that the proximity of the lines is greatest in the areas immediately adjacent to the sides of the hole.

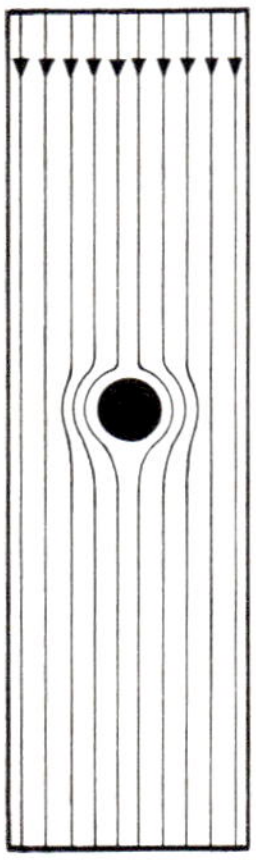

Fig. 7–15. While fastened to a bone, a plate is subjected to a bending loading. The loading causes tension to be produced on the surface of the plate. A fluid flow analogy which depicts the concentrations of the stress produced in the area of the hole through the plate is shown.

In areas distal to the hole, the lines are quite uniform in their spacing, indicating a uniform stress distribution. The effect of the stress concentration, then, is not simply a reduction of the area of the element, but rather it is a geometric discontinuity which induces a very local disturbance in the stress pattern. In the case of the hole in the plate, the stress in the area indicated adjacent to the hole is 3 times larger than the stress in those portions of the plate distal to the hole. An additional problem encountered with stress concentrations is their undesirable effect under suddenly applied loads. The normal mode of failure of an implant under a load of large magnitude, even though it is suddenly applied, is the appearance of a large permanent deformation in the device. If, however, a stress concentration is present, the material at the stress concentration site will deform first since it is most highly stressed. The remainder of the material does not reach yield stress, and therefore, a fracture passing through the point of stress concentration, as seen in Figure 3–13, results. Since the maximal applied load is reduced and the total deformation before fracture is likewise reduced, the amount of energy which an implant device may absorb under rapid loading conditions up to its point of failure is reduced.

Another subsystem commonly used in implant devices consists in screws. A discussion of the design characteristics of screws requires an understanding of the nomenclature associated with these elements.

Screw forms may be divided into two general classifications, machine screws and wood screws. The basic characteristic of a wood

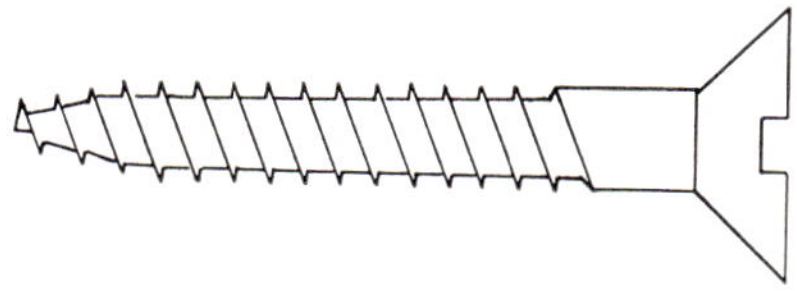

Fig. 7–16. Wood screw.

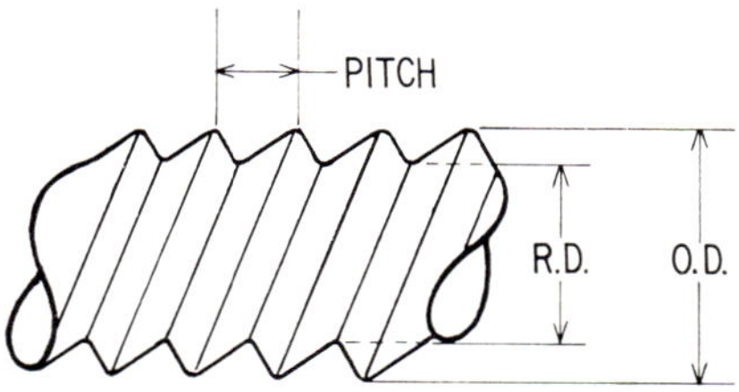

Fig. 7–17. A machine screw together with the commonly used nomenclature for describing the various portions of the screw is shown. Pitch is the distance between two successive threads. Outside diameter (O. D.) is the greatest diameter of the threaded portion. Root diameter (R.D.) refers to the minimum diameter created by the impression of the thread in the screw.

screw may be seen in examination of Figure 7–16. The front of the screw is tapered to a point and it is threaded up to this point. When inserting a wood-type screw, no hole need be drilled in the material. Rather, the screw itself forces its own pathway through the material into which it is inserted. In doing so, it crushes the material and compacts it around the threads. Thus their use is not compatible with hard, brittle materials such as cortical bone.

Machine screws have a blunt end which must be inserted into a threaded hole. This thread may be formed in one of two ways. A specially designed cutting tool, called a tap, may be passed through an undersized hole and in so doing, cut a threaded form into the hole. A second method is to include the tap in the forward portion of the machine screw. In such a case the screw itself taps the hole as it passes into the material. This type of machine screw is called a self-tapping screw.

There are four basic dimensions that must be considered in evaluating the performance characteristics of any type of thread. These quantities are: (1) root diameter; (2) outside diameter; (3) lead, and (4) pitch. In addition to these four, the thread shape is also a significant factor. The root diameter is the minimum inside diameter of the thread portion. This diameter determines the tensile strength of the screw since it determines the minimum cross-sectional area of the screw. The outside diameter, being a measure of the overall diameter of the thread, is one of two indications of what the pull-out strength of the screw will be when inserted in bone. (This will be discussed more fully in the next paragraph.) Pitch is the distance between thread forms. Thus, if there are 20 threads per linear inch on the screw, the pitch will be $1/20$ inch. Lead is the distance that the screw advances when it is turned through one revolution. The lead is actually the product of the pitch and the number of individual threads wrapped around the shank of the screw. In a single-thread screw, the lead is equal to the pitch. In a double-thread screw, the lead is twice the pitch. The advantages of multiple-thread screws are that for a given number of revolutions they will tighten more quickly.

The holding power, or the ability to resist stripping, of a machine screw is largely determined by the total cross-sectional area of material present at the root of the thread. This area is shown diagrammatically in Figure 7–18. If the material into which the screw is placed is much weaker than the screw material, the critical area is the root area of the tapped hole (Fig. 7–18B). The root area of the screw (Fig. 7–18A) will increase with increasing root diameter. The root area of the threaded hole will increase with increasing outside thread diameter.

Two thread forms which have substantially the same diameter and root area are shown in Figure 7–19. A third form is also shown in which the root area is significantly lower. This type of thread may be expected to have a lower pull-out strength than the others. Since the pull-out strength is dependent upon the cross-sectional area at

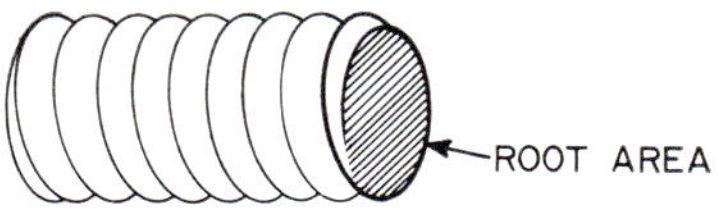

Fig. 7–18A. The root area of a machine screw.

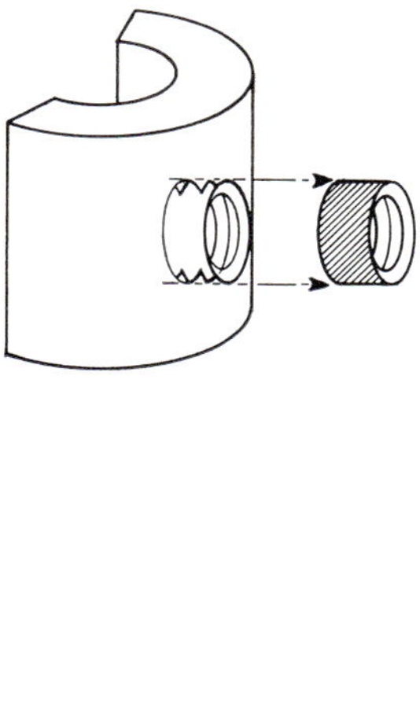

Fig. 7–18B. The root area of a tapped hole.

the base of the thread, neither the number of threads per inch nor pitch of the thread has a significant influence on this pull-out strength. For practical purposes, the only consideration in determining the pitch of a thread which will be satisfactory is that there must be a sufficient number of threads formed in the cortical section of the bone so that the accidental breakage or disruption of one or two threads does not cause a catastrophic weakening of the whole system. If 5 or 6 threads can be formed in the cortical section, a satisfactory holding power would be achieved even if one or perhaps two of the threads were malformed or broken.

The head design of a machine screw contributes greatly to the strength in a screw plate system. As illustrated in Figure 7–20, it is not always possible to obtain perfect alignment of all the screws inserted in a given implant device. Because of this, there exists a possibility that the head of the screw may not be perfectly aligned within the hole of the plate. If misalignment results in contact between the screw and plate, as shown in Figure 7–21, tightening of the screw will result in the production of a bending moment in the screw in the area immediately adjacent to the head of the screw. If the misalignment of the screw is such that the angle between the screw head and the surface of the plate is 2 degrees, tightening of the screw in the plate will produce bending stresses sufficient to cause yielding in the

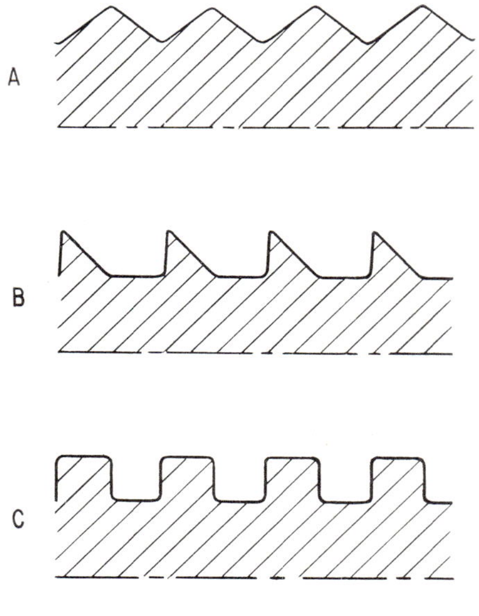

Fig. 7–19. Of the three thread forms shown, the first two, A and B, have essentially the same area of material at their root diameters. Thread form B has approximately ½ of the area of material at the root diameter.

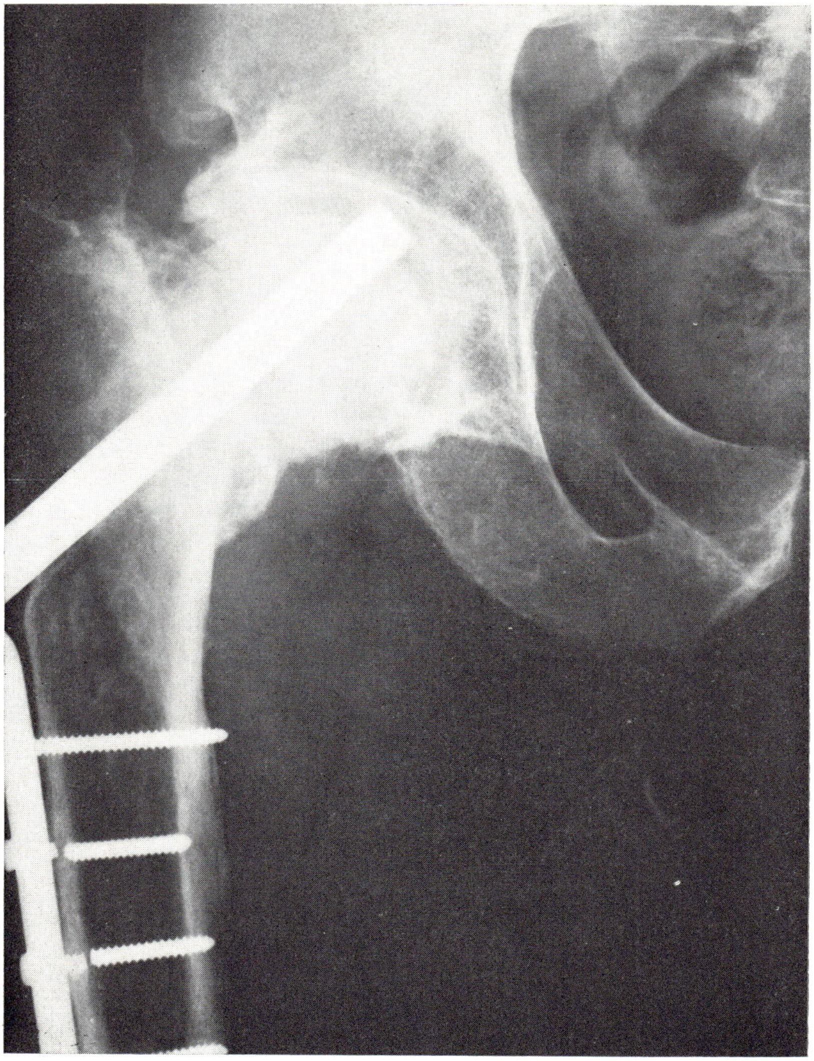

Fig. 7–20. A fracture of the femoral neck fixed with a nail plate. The screws were not aligned parallel to the holes in the nail plate so that breakage of the screw heads resulted.

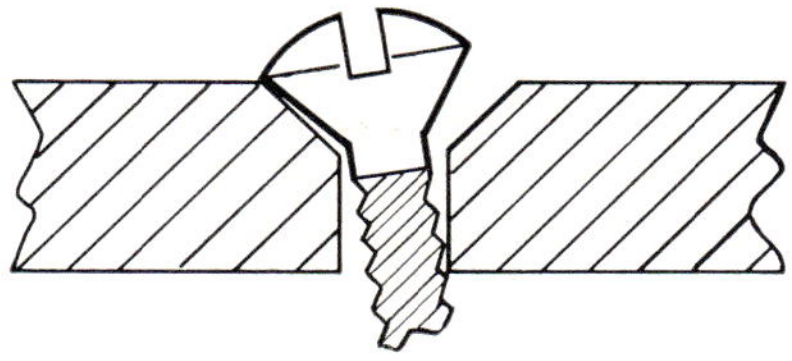

Fig. 7–21. If the screw is misaligned with the hole in the nail plate, a force on the head produces a bending moment.

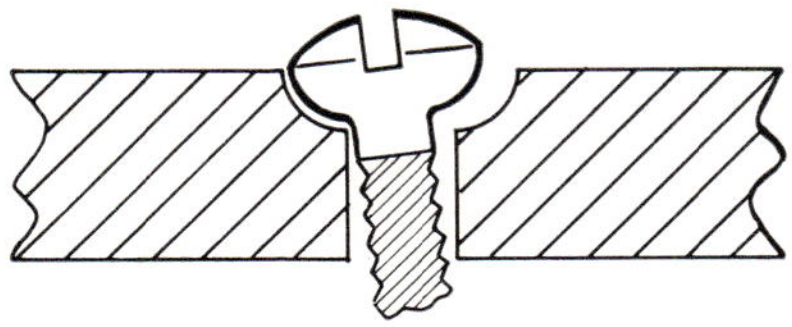

Fig. 7–22. A screw with a spherically shaped head has less chance of inducing bending loads at the plate.

local area of the head-neck junction. This stress is in addition to any other loadings imposed upon the screw either because of tension created in pulling the two cortical surfaces together or the shear caused by the torsional loading of the screwdriver in tightening the screw. In addition, in-service loading will take place. A design technique which is useful in avoiding this type of extraneously induced load is illustrated in Figure 7–22. The spherical geometry of this screw head and plate effectively reduces the possibility of premature failure in a screw because of misalignment.

Some of the principles developed in previous chapters and combined with the design concepts presented in this chapter may be utilized to develop an understanding of the behavior of internal fixation devices under the commonly encountered loading conditions. For purposes of this discussion, only transverse and oblique fractures in long bones will be considered.

A common method of providing fixation in an oblique fracture is by the use of a screw. There are several techniques utilized which differ in the direction and placement of the screw. Screws may be placed either perpendicular to the fracture, perpendicular to the long axis of the bone, or as a combination of the previous two. The bone may be subjected to various types of loading after fixation. Insertion of screws, which under the various loading configurations tend to keep the fracture surfaces together, seem to provide the most favorable climate for healing. Some configurations of fixation geometry and loading condition allow for distraction of the fracture surfaces or sliding of the fracture surfaces. In Table 7–1, a summary is presented which qualitatively describes the behavior at the fracture site when the bone is subjected to various permutations of loading condition and screw placement.

Tension is rarely observed on a bone after internal fixation and will be disregarded in the discussion. Compression is frequently encountered and two of the three configurations produce reasonable strength in compression. If the screws are placed perpendicular to the fracture line, some sliding and shortening may occur.

Bending is also a commonly encountered loading condition. To understand the effect of the screw placement on bending, it is necessary to consider the bending axis. *In vivo* this requires an understanding of the bending conditions on the bone. If the bending axis is parallel to the screw, there is minimal resistance to bending. If the bending axis is perpendicular to the screw placement, better stability is achieved. The best configuration is achieved by placing the screws in the combination position. If the screws are placed perpendicular to the long axis, the side with the compressive stresses will be the strongest.

Table 7–1.

SCREW GEOMETRY / LOADING MODE	PERPENDICULAR TO FRACTURE	PERPENDICULAR TO LONG AXIS	COMBINATION
COMPRESSION	SOME SLIDING DISPLACEMENT WILL OCCUR	STABLE; FRACTURE SITE IS IN COMPRESSION	STABLE; FRACTURE SITE IS IN COMPRESSION
TENSION	SOME DESTRUCTION DISPLACEMENT WILL OCCUR	MARKED DISPLACEMENT	SOME DESTRUCTION DISPLACEMENT WILL OCCUR
BENDING AXIS PARALLEL TO SCREWS 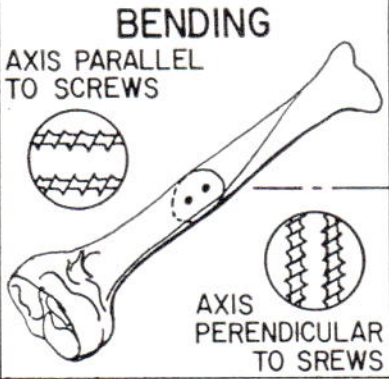	MINIMAL RESISTANCE	MINIMAL RESISTANCE	MINIMAL RESISTANCE
AXIS PERENDICULAR TO SREWS	MODERATE STABILITY	MODERATE STABILITY	BETTER STABILITY THAN EITHER OF THE OTHER GEOMETRIES
TORSION	MINIMAL STRENGTH	MINIMAL STRENGTH	MINIMAL STRENGTH

In the case of torsion, a commonly encountered loading condition, no significant strength can be achieved because all loading configurations subject the screws to bending at their insertion sites into the cortical bone. Since the screw is slender and not designed to withstand bending moments at its threaded insertion points, it will be displaced easily.

The design aspects of intramedullary rods which are critical in providing satisfactory internal fixation are the aspects that affect internal gripping, strength, and rigidity. First the nail must be designed so that it grips the inner aspects of the cortex both proximally and distally to the fracture site. Since the geometry of the inner aspect of the cortex varies throughout the length of the bone, the shape must be such that it fits the narrowest portion unless an efficient reaming device is used.

Utilizing a standard midshaft fracture as the example in all loading configurations, the system can only develop strength and rigidity in torsion if shearing stresses can be built up between the inner aspect of the cortex and the nail. Unless the shear stresses that normally develop in the cortical surfaces of the bone can be transferred from the inner aspect of the cortex proximal to the fracture site to the rod, the rod cannot sustain significant loading in torsion. For example, a round rod in a round hole cannot easily build up shear stresses at its surface, and slippage will occur. A diamond rod in a round hole may develop shear stresses if the vertices of the diamond have forcibly displaced bone tissue and develop an intimate contact. This will allow the torsional load to be transferred from the bone to the rod and thence to the proximal bone. Large contact pressures which allow the development of surface shear stresses may be developed if a springy rod is placed in the cavity. However, if the compressible rod has an open section such as a clover-leaf nail, and although it may be able to develop high shear stresses at the contact surface, its ability to carry torsional loads and provide rigid fixation is very small due to the open-section effect mentioned previously in Chapter 2.

In addition, there is a great difference in the torsional rigidity (per unit length) between the bone and the open-section nail. The torsional rigidity (G times J) of an average femur is in the region of 1.5×10^5 to 2.5×10^5 kg•cm^2. For a clover-leaf nail of appropriate size the torsional rigidity is 1.6×10^3 kg•cm^2. The open-section nail is about $1/100$ as rigid as the bone. A diamond-section nail of appropriate size has a torsional rigidity of 1.2×10^4 kg•cm^2, while a Schneider nail produces 1.5×10^4 kg•cm^2. These closed-section nails are therefore 10 times as rigid as the open-section nails, and about one-tenth as rigid as the bone they are supporting.

In bending, it is easy to transmit the load to the rod so that the bending strength of the system will be that of the rod itself. The nature of the cross section and, in particular, the area moment of inertia will determine the strength of the system in any particular bending direction.

For the diamond nail, the bending strength, which is defined as the moment required to produce permanent deformation, varies from 210 kg•cm for a bending axis through the long diagonal to 250 kg•cm for a bending axis through the short diagonal. The bending strength of the open-section clover-leaf nail varies from 290 to 350 kg•cm depending upon which axis is chosen. The Schneider nail has a bending strength of between 285 and 400 kg•cm. These figures are to be compared with the bending strength of 780 to 880 kg•cm for the bone in which these nails would be used.

In compression loading, the bone itself takes up the load: the rod only maintains the geometry. In considering the biomechanics of plate fixation, a similar transverse midshaft fracture is utilized. The principles developed in Chapter 2 can be applied to an analysis of plate fixation. In compression loading with a transverse fracture, the bone takes up the load, the plate merely serves to maintain alignment.

The bending strength of the fixation, that is, the greatest bending moment which can be sustained by the plated bone, is dependent upon the bending axis and the sense of the bending moment. If the sense of the bending moment is such as to put the plate in compression and thus try to put the opposite side of the bone in tension, the system will not sustain great loads (a in Fig. 7–23). The strength of the system will be simply the bending strength of the plate in its weakest dimension, and its rigidity is approximately 4.8×10^3 kg•cm^2 for an average tibia. If the sense of the bending moment is such as to place the plate under tension and the opposite surface of the bone under compression, as shown in b in Figure 7–23, a strong and rigid configuration will result. This is due to the fact that the compressive stresses required in bending may be transmitted across the fracture. This configuration is more than 100 times as rigid as the previous case. If the bending axis is at right angles to the plate, the screws will be subjected to a combination of bending and torsional loading, and although the plate is being bent around a stronger axis than in a in Figure 7–23, the entire configuration may fail because of the torsion and bending loading of the screws. Thus, this system falls somewhere between the ultimate bending moments of the systems illustrated in a and b of Figure 7–23.

If a second plate is added at right angles to the first plate, the system is stronger in bending in all axes. The two plates will not permit the development of the situation as shown in c in Figure 7–23.

The final loading configuration is that of torsion. The transverse fracture prevents the bone from transmitting any torsion because only minimal shear stresses can be built up at the interdigitating surfaces. All the torsion must be transmitted through the plate. Comparing the thin flat plate to the round tubular bone, it is apparent from an examination of the polar moment of inertia that the plate's polar moment of inertia will only be a fraction of that of the round tubular bone. However, the torsional rigidity is the product of the polar moment of inertia and the shear modulus. Therefore, a heavy steel plate ⅛ of an inch thick and ⅝ of an inch wide with a 1½ inch spacing between screw holes will have a torsional rigidity of 54 inch pounds per degree. This compares favorably to the 45 inch pounds per degree of an intact tibia over the same 1½ inch length. The

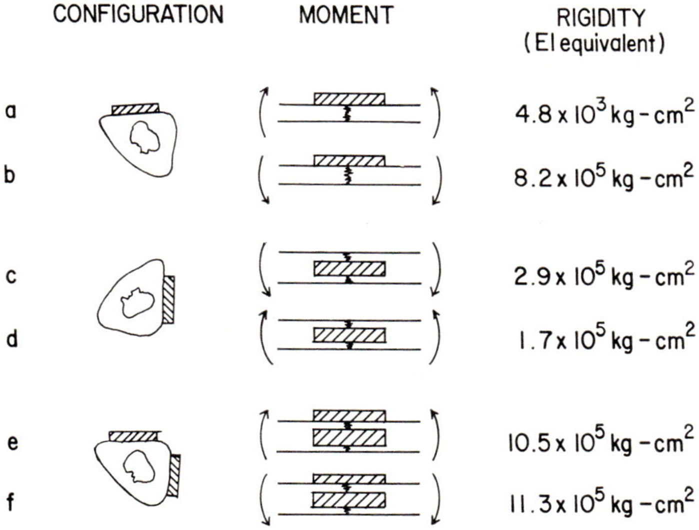

Fig. 7–23. Bending rigidity of various plate configurations.

actual torsional rigidity of the fixation will be less than the ideal rigidity because of bending in the screws. The effect of adding a second plate at right angles to the system will be simply that of doubling the rigidity. The design consideration presented above may be a guide in the development of internal fixation systems and should be combined with the knowledge of normal loading configurations in order that a successful fixation system may be developed.

Bibliography

This Bibliography is intended only as a guide to further reading. Whenever possible, articles have been chosen which contain comprehensive bibliographies.

Chapter 1

1. BASMAJIAN, J. V.: *Muscles Alive.* Baltimore, Williams & Wilkins, 2nd Ed., 1967.
2. BECHTOL, C. O.: Surgical application of muscle physiology, A.A.O.S. Instructional Course Lecture, Ann Arbor, J. W. Edwards, *XII*:161, 1955.
3. BOURNE, G. H.: *The Structures and Function of Muscle.* New York, Academic Press, 1960.
4. BUCHTHAL, F., SVENSMARK, O., and ROSENFALCK, P.: Mechanical and chemical events in muscle contraction, Physiol. Rev. *36*: 503, 1956.
5. CLOSE, J. R.: *Motor Function in the Lower Extremity.* Springfield, Ill., Charles C Thomas, 1964.
6. DEMPSTER, W. T.: Space Requirements of the Seated Operator; Geometrical, Kinematic, and Mechanical Aspects of the Body with Special Reference to the Limbs. WADC Technical Report B. No. 55-159, July, 1955.
7. DEMPSTER, W. T.: "Free-Body Diagrams as an Approach to the Mechanics of Human Posture and Motion," *Biomechanical Studies of the Musculo-Skeletal System.* Edited by F. G. Evans, Springfield, Ill., Charles C Thomas, 1961.
8. DRILLIS, R., and CONTINI, R.: *Body Segment Parameters.* Technical Report No. 1166 03, New York University School of Engineering and Science, University Heights, New York, 1966.
9. ELFTMAN, H.: Biomechanics of muscle, J. Bone Joint Surg. *48A*: 363, 1966.

10. INMAN, V. T.: Functional aspects of the abductor muscles of the hip, J. Bone Joint Surg. *29*:607, 1947.
11. KENEDI, R. M.: *Biomechanics and Related Bio-Engineering Topics.* New York, Pergamon Press, 1965.
12. OMER, G. E., JR., and VOGEL, J. A.: Determination of physiological length of a reconstructed muscle—Tendon unit through muscle stimulation, J. Bone Joint Surg. *47A*:304, 1965.
13. RODAHL, K., and HORVATH, S. M.: *Muscle as a Tissue.* New York, McGraw-Hill, 1962.
14. RYDELL, N.: "Intravital Measurements of Forces Acting on Hip-Joint." *Studies on the Anatomy and Function of Bones and Joints.* Edited by F. G. Evans, Heidelberg, Springer-Verlag, 1966.
15. SHAMES, I. H.: *Engineering Mechanics: Statics.* Englewood Cliffs, New Jersey, Prentice-Hall, Inc., 1959.
16. SINGER, F. L.: *Engineering Mechanics.* New York, Harper & Row, 2d Ed., 1954.
17. WILLIAMS, M., and LISSNER, H. R.: *Biomechanics of Human Motion.* Philadelphia, W. B. Saunders, 1962.
18. ZAREK, J. M.: "Dynamic Considerations in Load Bearing Bones with Special Reference to Osteosynthesis and Articular Cartilage," *Studies on the Anatomy and Function of Bone and Joints.* Edited by F. G. Evans, New York, Springer-Verlag, 1966.

Chapter 2

19. AMPRINO, R.: "Microhardness Testing as a Means of Analysis of Bone Tissue Biophysical Properties," *Biomechanical Studies of the Musculo-Skeletal System.* Edited by F. G. Evans, Springfield, Ill., Charles C Thomas, 1961.
20. ASCENZI, A., BONUCCI, E., and CHECCUCCI, A.: "The Tensile Properties of Single Osteons Studied Using a Microwave Extensimeter," *Studies on the Anatomy and Function of Bone and Joints.* Edited by F. G. Evans, New York, Springer-Verlag, 1966.
21. EVANS, F. G.: *Stress and Strain in Bones.* Springfield, Ill., Charles C Thomas, 1956.
22. EVANS, F. G., and BANG, S.: "Physical and Histological Differences Between Human Fibular and Femoral Compact Bone," *Studies on the Anatomy and Function of Bone and Joints.* Edited by F. G. Evans, New York, Springer-Verlag, 1966.

23. EVANS, F. G., and KING, A.: "Regional Differences in Some Physical Properties of Human Spongy Bone," *Biomechanical Studies of the Musculo-Skeletal System.* Edited by F. G. Evans, Springfield, Ill., Charles C Thomas, 1961.

24. FRANKEL, V. H., and BURSTEIN, A. H.: "Load Capacity of Tubular Bone," *Biomechanics and Related Bio-Engineering Topics.* Edited by R. M. Kenedi, New York, Pergamon Press, 1965.

25. FRANKEL, V. H., and BURSTEIN, A. H.: The biomechanics of refracture of bone, Clin. Orthop. *60*:221, 1968.

26. MATHER, B. S.: Correlation between strength and other properties of long bones, J. Trauma *7*:633, 1967.

27. NACHEMSON, A.: Lumbar intradiscal pressure: Experimental studies on postmortem material, Acta Orthop. Scand. Suppl. *43*: 1960.

28. ROTHMAN, R. H.: "Electrical and Mechanical Principles in Bone Biodynamics," *Engineering in the Practice of Medicine.* Edited by B. L. Segal and D. G. Kilpatrick, Baltimore, Williams & Wilkins, 1967.

29. SEELY, F. B., and SMITH, J. O.: *Resistance of Materials.* New York, John Wiley & Sons, Inc., 4th Ed., 1956.

30. SINGER, F. L.: *Strength of Materials.* New York, Harper & Row, 2d Ed., 1951.

31. TIMOSHENKO, S.: *Strength of Materials. Part I, Elementary Theory and Problems.* New York, D. Van Nostrand Company, 3d Ed., 1955.

32. TIMOSHENKO, S., and MACCULLOUGH G. H.: *Elements of Strength of Materials.* New York, D. Van Nostrand Co., 3d Ed., 1949.

33. VIIDIK, A.: "Biomechanics and Functional Adaption of Tendons and Joint Ligaments," *Studies on the Anatomy and Function of Bone and Joints.* Edited by F. G. Evans, New York, Springer-Verlag, 1966.

Chapter 3

34a. HICKMANN, KENNETH: Rheological Behavior of Tissues Subjected to External Pressure, Ph.D. Thesis, Case Institute of Technology, School of Engineering, 1965.

34b. HOUSNER, G. W., and VREELAND, T., JR.: *The Analysis of Stress and Deformation.* New York, Macmillan Company, 1966.

34c. SHAMES, I. H.: *Engineering Mechanics: Statics.* Englewood Cliffs, New Jersey, Prentice-Hall, Inc., 1959.

34d. SINGER, F. L.: *Strength of Materials.* New York, Harper & Row, 2nd Ed., 1951.

34e. SINGER, F. L.: *Engineering Mechanics.* New York, Harper & Row, 2nd Ed., 1954.

34f. TIMOSHENKO, S.: *Strength of Materials. Part I, Elementary Theory and Problems.* New York, D. Van Nostrand Company, 3rd Ed., 1955.

34g. TIMOSHENKO, S., and MACCULLOUGH, G. H.: *Elements of Strength of Materials.* New York, D. Van Nostrand Company, 3rd Ed., 1949.

Chapter 4

35. BURSTEIN, A. H., and FRANKEL, V. H.: The viscoelastic properties of some biological materials, Ann. N. Y. Acad. Sci. *146*: 158, 1968.

36. CAMOSSO, M. E., and MAROTTI, G.: The mechanical behavior of articular cartilage under compressive stress, J. Bone Joint Surg. *44A*:699, 1962.

37. GALANTE, J. O.: Tensile properties of the human lumbar annulus fibrosis, Acta Orthop. Scand. Suppl. *100,* 1967.

38. HOUSNER, G. W., and VREELAND, T., JR.: *The Analysis of Stress and Deformation.* New York, Macmillan Company, 1966.

39. MATTHEWS, L. S., and ELLIS, D.: Viscoelastic properties of cat tendon, J. Biomechanics *1*:65, 1968.

40. SEDLIN, E.: A rheological model for cortical bone, Acta Orthop. Scand. Suppl. *83,* 1965.

Chapter 5

41. HALL, A. S., JR.: *Kinematics and Linkage Design.* Englewood Cliffs, New Jersey, Prentice-Hall, Inc., 1961.

42. HARRISBERGER, L.: *Mechanization of Motion.* New York, John Wiley & Sons, Inc., 1961.

43. HINKLE, R. T.: *Kinematics of Motion.* Englewood Cliffs, New Jersey, Prentice-Hall, Inc., 2d. Ed., 1960.

44. SAUNDERS, J. B. DEC.M., INMAN, V. T., and EBERHART, H. D.: The major determinants in normal and pathologic gait, J. Bone Joint Surg. *35A*:543, 1953.

45. SLOANE, A.: *Engineering Kinematics.* New York, Macmillan Company, 1954.

46. STEINDLER, A.: *Kinesiology of the Human Body.* Springfield, Ill., Charles C Thomas, 1955.

Chapter 6

47. BARNETT, C. H., and COBBOLD, A. F.: Lubrication within living joints, J. Bone Joint Surg. *44B*:662, 1962.
48. BARNETT, C. H. *et al.*: *Synovial Joints: Their Structure and Mechanics.* Springfield, Ill., Charles C Thomas, 1961.
49. DINTENFASS, L.: Rheology of complex fluids and some observations on joint lubrication, Fed. Proc. *25*:1054, 1966.
50. MCCUTCHEON, C. W.: "Physiological lubrication," *Lubrication and Wear in Living and Artificial Human Joints.* Symposium. The Institution of Mechanical Engineers, London, April, 1967.
51. SHIGLEY, J. E.: *Mechanical Engineering Design.* New York, McGraw-Hill, 1963.
52. TANNER, R. I.: An alternative mechanism for the lubrication of synovial joints, Phys. Med. Biol. *11*:119, 1966.
53. TRUMPLER, P. R.: *Design of Film Bearings.* New York, Macmillan Co., 1966.
54. WHITE, R. K.: The clinical applications of the rheology of synovial fluid, Clin. Orthop. *29*:213, 1863.

Chapter 7

55. ALLEN, W. C., PIOTROWSKI, G., BURSTEIN, A. H., and FRANKEL, V. H.: Biomechanical principles of intramedullary fixation, Clin. Orthop. *60*:13, 1958.
56. BECHTOL, C. O.: Engineering principles as applied to orthopaedic surgery, A.A.O.S. Instructional Course Lecture, Ann Arbor, J. W. Edwards, *9*:257, 1952.
57. BECHTOL, C. O., FERGUSON, A. B., JR., and LAING, P. G.: *Metals and Engineering in Bone and Joint Surgery.* Baltimore, Williams & Wilkins, 1959.
58. CAPENER, N.: Surgical implants, J. Bone Joint Surg. *47B*:3, 1965.
59. COHEN, J.: Assay of foreign-body reaction, J. Bone Joint Surg., *41A*:152, 1959.
60. COHEN, J.: Tissue reactions to metals—The influence of surface finish, J. Bone Joint Surg. *43A*:687, 1961.
61. COHEN, J.: Corrosion testing of orthopaedic implants, J. Bone Joint Surg. *44A*:307, 1962.
62. COHEN, J.: Failure in performance of surgical implants, J. Bone Joint Surg. *46A*:416, 1964.
63. FERGUSON, A. B., JR., LAING, P. G., and HODGE, E. S.: The ionization of metal implants in living tissues, J. Bone Joint Surg., *42A*: 77, 1960.

64. Frankel, V. H.: *The Femoral Neck.* Springfield, Ill., Charles C Thomas, 1960.

65. Frankel, V. H., and Burstein, A.: "The Design of Orthopaedic Prostheses," *Engineering in the Practice of Medicine.* Edited by B. L. Segal and D. G. Kilpatrick, Baltimore, Williams & Wilkins, 1967.

66. Gillis, T.: *Modern Trends in Surgical Materials.* London, Butterworth, 1958.

67. Hicks, J. H., and Cater, W. H.: Minor reactions due to modern metal, J. Bone Joint Surg. *44B*:122, 1962.

68. Laing, P. G.: The significance of metallic transfer in the corrosion of orthopaedic screws, J. Bone Joint Surg. *40A*:853, 1958.

69. Lindahl, O.: Rigidity of immobilization of oblique fractures, Acta Orthop. Scand. *35*:39, 1964.

70. Lindahl, O.: The rigidity of fracture immobilization with plates, Acta Orthop. Scand. *38*:101, 1967.

71. Siegel, M. J., Maleev, V. L., and Hartman, J. B.: *Mechanical Design of Machines.* Scranton, Pennsylvania, International Textbook Company, 1965.

72. Spotts, M. G.: *Mechanical Design Analysis.* Englewood Cliffs, New Jersey, Prentice-Hall, Inc., 1964.

73. Wickstrom, J.: Symposium: Surgical mechanics of the internal fixation of fractures, J. Bone Joint Surg. *46A*:397, 1964.

Index